# 50 Policies and Plans for Outpatient Services

# 50 Policies and Plans for Outpatient Services

Carole Guinane, RN, MBA and Joseph Venturelli

CRC Press
Taylor & Francis Group
Boca Raton  London  New York

CRC Press is an imprint of the
Taylor & Francis Group, an **informa** business

A PRODUCTIVITY PRESS BOOK

CRC Press
Taylor & Francis Group
6000 Broken Sound Parkway NW, Suite 300
Boca Raton, FL 33487-2742

© 2012 by Taylor & Francis Group, LLC
CRC Press is an imprint of Taylor & Francis Group, an Informa business

Printed in the United States of America on acid-free paper
Version Date: 20111031

International Standard Book Number: 978-1-4398-6842-3 (Paperback)

---

**Library of Congress Cataloging-in-Publication Data**

---

Guinane, Carole S.
    50 policies and plans for outpatient services / Carole Guinane, Joseph Venturelli.
        p. ; cm.
    Fifty policies and plans for outpatient services
    Includes bibliographical references and index.
    ISBN 978-1-4398-6842-3 (pbk. : alk. paper)
    I. Venturelli, Joseph. II. Title. III. Title: Fifty policies and plans for outpatient services.
    [DNLM: 1. Ambulatory Care--organization & administration. 2. Ambulatory Care--standards. 3. Health Policy. 4. Quality Assurance, Health Care. 5. Safety Management. WB 101]

362.12068--dc23

2011043032

---

**Visit the Taylor & Francis Web site at**
**http://www.taylorandfrancis.com**

**and the CRC Press Web site at**
**http://www.crcpress.com**

# Contents

# Preface

"We can't run. We can't pass. We can't stop the run. We can't stop the pass. We can't kick. Other than that, we're just not a very good football team right now."

**—Bruce Coslet**

## OVERVIEW

The quote selected above is pertinent to the overall intent of this book. If employees, physicians, allied health professionals and contract workers can't run, can't pass, can't kick, can't stop the run or stop the pass, then we aren't a very good healthcare team. Translated, that means we wouldn't know what to do, when to do it, how to do it and why we are doing it for our patients and for each other.

Documents play such an important role in healthcare. They are the bread for our butter. We can't live without them and yet on many days we can't stand the sight of one more piece of paper on our desk. However, policies, procedures and plans, in part, help us know what to do, when to do it and why are doing what we are doing in the healthcare setting. They help reduce and eliminate variation, which ultimately reduces and eliminates errors.

Documents of this nature are great teaching tools. They support our mission, help us achieve our vision and support our values day-to-day. They are fundamental building blocks for the culture of an organization. If documents are poorly researched, written and executed, the results can and will lead to great losses in more ways than one.

The documents included in this book are not intended to implement as is—they are templates. The template definition that we adopted is as follows: *a model or standard for making comparisons*. So keep that in mind—do not use these documents as is; that is not the intent.

Know your state licensing requirements as well as other governing agency requirements. Make sure you have covered those basics. Then, most importantly, the documents should say what you want them to say in order to create the flow and process to reflect your organization intricacies and unique qualities. They all need to work with one another as well, without any conflict or confusion.

## BOOK OVERVIEW

This book is made up of three parts. We lead off with a "how to" chapter, guiding you with information about how to format the documents and how to make them your own. We then deliver the 50 documents, which are presented in alphabetical order according to the title.

Finally, a list of resources to help you with your journey is included at the end of the book, containing information about a recommended companion book, *Improving Quality in the Outpatient Setting* which includes an appendix that provides information about all state agencies. If that is a need for you, this companion book provides that resource.

## PURPOSEFUL OMISSIONS

As noted earlier, the documents are templates. We did not check them against every local, state and national law, state licensing requirement or regulatory agency standards. They are basic and they cover the basics. To create one policy that fits every state and organization in this country would be overwhelming to any user, as all state laws and licensing requirements are different.

While there are 50 documents, these alone do not supply you with a complete policy and procedure manual. It's a beginning. For more information regarding what other policies, procedures and plans to incorporate, please review the section on documents in the book, *Improving Quality in the Outpatient Setting*. A detailed policy, procedure and plan list can be found.

# Acknowledgements

Thank you, Kristine Mednansky, Senior Editor at CRC Press, A Taylor and Francis Group. You are the greatest coach and we appreciate your guidance and support from beginning to end. We couldn't have done it without you.

A hearty thanks to all that contribute to knowledge enhancement through document creation. These professionals enrich the lives of so many people in the healthcare world each and every day.

We also say thank you to the healthcare librarians. As the finders and keepers of the nuggets of information that contribute to document formation, your talents are most welcome.

## FROM CAROLE

I have been dubbed the document queen on many occasions. At first, I didn't quite know how to respond, but in time, I came to value the title and honor. I like to read, to ask questions and to think about different ways to do things if it improves a patient's journey. Plus, I like the challenge of telling a story through these documents. Writing guidelines, protocols, pathways and such are tools that help get our care processes standardized while policies, procedures and plans are the back-up documents that give us the rationale, vision and theory behind those tools. They are all necessary and useful.

I can personally attest to using keenly written policies, procedures and plans as references and tools when making clinical and administrative decisions. I am grateful for reference availability that's easily accessible as a great deal of time and effort is saved. I use documents every day of my life to help me succeed with small and large tasks. Whether it's the detailed actions required for approving payroll or the necessary steps I must take when completing a root cause analysis, all are equally appreciated.

A special thanks to my co-author and wonderful friend, Joseph, as he is a technical guru. He has a special knack with document manipulation and presentation. When working on a project such as this, those skills are invaluable.

This book is dedicated to my grandchildren, Darcy Elizabeth, Carolyn Rose, William Alexander and Rowan Grace, with love; and to my daughter, Carissa, for always being there for me. A special thanks to my husband Tom for tolerating the long hours it takes to complete a book such as this.

## FROM JOSEPH

Back in my early days of health care I had no idea the long term impact it would have on my life. My first real opportunity in health care was when I was hired as a system administrator for a hospital in Charlotte North Carolina. After being born and raised in the boroughs of New York this was an unlikely place for me to start my career in health care.

Not until many, many years later did I understand the importance that technology, standardization and documentation would play in delivering true patient focused care.

When I found myself on the receiving end of some really poor care that nearly cost me my life it became clear to me that patient advocacy and anything I could do to that end was a journey worth the effort.

I want to thank my dear friend Carole who asked me to join her in this project. She has inspired me and pushed me when I needed it and has been a true partner in this project.

I also want to thank my wife Barbara for her seemingly endless supply of love and support. She has seen me through every great success and been by my side through every stumble. The success of our careers has taken time from away from us yet our love was never in question.

This book is dedicated to my wife Barbara, and my children Michael and ToniAnn.

# The Authors

## CAROLE GUINANE RN, MBA

Carole's quality and leadership journey began in 1989 as a senior leader/Vice President at Parkview Episcopal Medical Center in Pueblo, Colorado. Parkview's success story was published in 1992 by The Joint Commission, with the forward of the book written by Donald M. Berwick, MD. The book, *Striving for Improvement: Six Hospitals in Search of Quality* shared the process, methods and rewards that our leadership team, employees and physicians experienced. It was magical. Applying quality principles to clinical processes was new to healthcare at the time, but groundbreaking results occurred. Carole took the lessons learned from those early days and continues to grow her knowledge base for operational and clinical improvement application.

Carole has worked as a Chief Clinical and Compliance Officer for an Ambulatory Surgery Center company, a Vice President of Medical Staff Services and Quality for a healthcare system, A Vice President for Applied Business Science and Education for a specialty hospital and healthcare system, a Consultant/Clinical Improvement Director for a Center for Continuous Improvement and Innovation, and, a Vice President for Ambulatory Clinical Improvement/ASC Clinical Operations for an integrated healthcare system. She has had the pleasure of building and growing quality and clinical operations programs for large healthcare systems, small and rural hospitals, ambulatory surgery centers, insurance companies and ambulatory entities. Carole is a trained Six Sigma Black Belt. She has published books and journal articles on clinical pathways, quality tools, Six Sigma, clinical operations, and consumer driven healthcare.

## JOSEPH VENTURELLI

Joseph earned his degree in design from the School of Visual Arts in New York City. His debut into healthcare began as a system administrator, concentrating on the technical oversight of information systems at Presbyterian Hospital in Charlotte, North Carolina in 1990. Early on Joseph trained and became certified as a system engineer and certified trainer. He has been responsible for leading teams of technologists including infrastructure, web services, data center operations, call centers, disaster recovery planning and help desk services, and has managed scores of system implementations over the past two decades.

Joseph co-led the creation and implementation of an electronic medical record system, which included full financial, transcription and scheduling integration. Physicians were given office access to the scheduling and clinical documentation platforms. This integration strategy created immediate medical record completion, eliminating the need for backend inspection and rework resources. Joseph has consistently delivered efficiencies through standardization, recruiting and retaining key talent and launching enthusiastic customer service programs for a variety of professionals, patients and vendors.

Joseph has been published in numerous technical journals and industry magazines and has authored several books including one on patient advocacy. A seasoned executive, Joseph has worked as the Chief Executive Officer for a Southeast Consulting firm, a Chief Information Officer for an Ambulatory Surgery Center company as well as the Chief Information Officer for a Midwest county hospital system.

# 1 How to Use This Book

Be faithful in small things because it is in them that
your strength lies.

—**Mother Teresa**

## THINK BEFORE YOU WRITE

When writing policies, procedures, plans, or algorithms, an
important fact to remember is to make them yours. Taking
any document and using it "as is" does not adequately reflect
what you want to have happen in your organization. These
documents contribute significantly to an organization's culture and sense of self. They guide us in our decision making.
They help us problem solve. Policies provide guidelines for
acceptable and unacceptable behaviors. Plans send us in the
right direction as they navigate pathways for us.

Documents should never be taken lightly as they are pow-
erful tools that give meaning to the mission, vision, and values
of your organization. It's paying attention to detail, and it's the
small things that really do matter, ultimately contributing to
creating and sustaining a strong organization.

More often than not, policies, plans, and key documents
are relegated to employees in an organization that gets it
done—meaning it's written, it's put in a binder, it's signed
off, and it goes on a shelf. Staff and physicians are trained
briefly on the book so they know how to reference it for surveys. They are given cheat sheets to enable them to answer
surveyor questions correctly.

While this thinking may meet, at a minimum, the regulatory and licensing requirements, the documents will go unused
until a surveyor shows up and leaders point to the book, delivering a convincing show and tell. When this scenario occurs,
the customers of these documents are the surveyors. That is
unfortunate. Ownership, accountability, and full implementation of key documents do not exist.

In essence, documents should help employees and other
stakeholders with the day-to-day duties of their job. Many
industries know the value of checklists and detailed procedures, as that's what sets the tone for safety, competency
training, as well as consistency. For example, an airplane will
not take off if the staff has not completed their safety checklists and followed their policies and procedures explicitly.

Leaders must be aware that organizations such as The Joint
Commission (TJC) are not the only regulatory or accrediting
body that can visit or survey. When policies reference TJC's
standards throughout, it shows that federal, state, and local
requirements are not included. For example, Medicare, the
Drug Enforcement Agency, The Department of Labor, the
Center for Disease Control, and state health departments can
fine an organization, shut an organization down, or place the
organization in immediate jeopardy if their regulations are

not met. ALL licensing and regulatory bodies must be considered when writing and implementing documents.

In healthcare, we can't afford to be a show-and-tell industry.
Patients' lives depend on us. As caregivers and healthcare professionals, we have to be great at what we are doing. Mistakes
cost lives and can forever alter the paths of those that are
touched. Our patients, staff, and physicians deserve tools and
methods to help cut through complexity whether through
detailed checklists, reference guides, and, yes, policies, procedures, and plans.

## ASSESSMENT TOOLS

Table 1 provides a prioritization matrix tool to help you determine if a policy, plan, or other document is meeting the needs
of the organization. Table 2 defines the scoring guideline for
the matrix. The scores noted in Table 1 stemmed from the
results of a leadership team's review of the policies noted in
the matrix. All of the policies scored high, as they directly
impact the key elements of this surgery center, therefore all
are critical for operations and patient care.

Pulling together your leadership team to contemplate the
answers and to think through what you want to say and do is
compelling. A matrix of this type also helps with prioritization. Take some time to create a matrix for your outpatient
program—use it as presented or edit it to enrich its value for
your leadership needs.

## DOCUMENT FORMAT

Figure 1 displays the format used for all documents in this
book. We created this design for the following reasons:

1. Errors are reduced when users know what to expect.
   That is why the format is the same for each and
   every document.
2. Labeling through logo/organization name use makes
   it yours.
3. It's easy to find and replace identifiers.
4. It works.

## VARIABLES AND HOW TO USE THEM

The collection of documents set forth in this book is included
on a CD-ROM disc for quick access and editing. To make
this user friendly, each document features a set of predefined
variables to simplify the customization process.

The documents are provided in Microsoft Word® format
so that you can use the search and replace function to change
the variables in the document to match your organization.

**TABLE 1**
**Prioritization Matrix**

| Document | Impact on Users | Impact on Operations | Impact on Goodwill | Impact on Licensure, Regulatory, Accreditation, Legal | Impact on Patient | Total Points |
|---|---|---|---|---|---|---|
| Competency Assessment Policy | 12 | 12 | 10 | 12 | 12 | 58 |
| Fall Prevention Policy | 9 | 11 | 9 | 12 | 12 | 53 |
| Hand Hygiene Policy | 12 | 11 | 10 | 12 | 12 | 57 |
| History and Physical Assessment Policy | 10 | 12 | 9 | 12 | 12 | 55 |
| Total Points | 43 | 46 | 38 | 48 | 48 | 223 |

**TABLE 2**
**Scoring Guide for Prioritization Matrix**

| Scoring System | Impact on Users | Impact on Operations | Impact on Goodwill | Impact on Licensure, Regulatory, Accreditation, Legal | Impact on Patient |
|---|---|---|---|---|---|
| Strong: 9 to 12 points | Affects more than 50% of users | Impacts core business functions and strategy | Would impact community and company positively or negatively | Would place the company in immediate jeopardy if not done | Would impact patient, family and significant others positively or negatively |
| Medium: 5 to 8 points | Affects greater than 10% but not more than 49% of users | Impacts non-core business functions and strategy | Would impact company positively or negatively | Would impact the company positively or negatively | Would impact patient positively or negatively |
| Weak: 1 to 4 points | Affects less than 10% of users | Impacts normal completion of work/tasks, but the work/tasks are not difficult to complete | Business unit/ department/individual is impacted positively or negatively | Recommended but not required | Recommended but not necessary for the patient |
| Not Applicable: zero points | Affects a single user | Impacts non-business related processes | No impact on Goodwill | No impact on licensure, regulatory, accreditation, and/or legal | Would not reach patient, family members, and significant others |

Each variable is surrounded by pound signs "#" so they can easily be replaced.

For example, if your organization is located in the state of Colorado and you wanted to make adjustments to the policy templates to reflect the name of your state, you would simply search the documents for the variable #STATE# and replace it with "Colorado".

The variables included in the documents and their definitions are as follows:

#POLNUM#     Use this variable to search and replace the Structured Policy Number

#LOC#          Use this variable to search and replace the Location Name or Code

#ORIGDEPT#   Use this variable to search and replace the Originating Department Name

#EFFDATE#     Use this variable to search and replace the Effective Date of the policy

#EXPDATE#     Use this variable to search and replace the Expiration Date of the policy

#ORIGDATE#    Use this variable to search and replace the Origination Date of the policy

#APPROVER#    Use this variable to search and replace the Name of the person approving the Policy

#APPRTITLE#   Use this variable to search and replace the Title of the Person Approving the policy

#APPRDATE#    Use this variable to search and replace the Date of the policy Approval

#STATE#        Use this variable to define the State for the organization's licensing and other requirements, including location

#CITY#          Use this variable to define the City's location

In addition to the CD-ROM containing all the policies in this book as individual documents, it will also include a single Microsoft Word document with all the policies included in one file so that you are able to initiate a global replace on all the variables outlined above.

**Policy and Procedures**

Policy#: #POLNUM#
Location: #LOC#
Originating Department: #ORIGDEPT#
Effective Date: #EFFDATE#
Expiration Date: #EXPDATE#

## TITLE: TITLE GOES HERE

### POLICY STATEMENT:

"Lorem ipsum dolor sit amet, consectetur adipisicing elit, sed do eiusmod tempor incididunt ut labore et dolore magna aliqua. Ut enim ad minim veniam, quis nostrud exercitation ullamco laboris nisi ut aliquip ex ea commodo consequat. Duis aute irure dolor in reprehenderit in voluptate velit esse cillum dolore eu fugiat nulla pariatur. Excepteur sint occaecat cupidatat non proident, sunt in culpa qui officia deserunt mollit anim id est laborum."

### INTENT AND SCOPE:

"Lorem ipsum dolor sit amet, consectetur adipisicing elit, sed do eiusmod tempor incididunt ut labore et dolore magna aliqua. Ut enim ad minim veniam, quis nostrud exercitation ullamco laboris nisi ut aliquip ex ea commodo consequat. Duis aute irure dolor in reprehenderit in voluptate velit esse cillum dolore eu fugiat nulla pariatur. Excepteur sint occaecat cupidatat non proident, sunt in culpa qui officia deserunt mollit anim id est laborum."

### DEFINITIONS:

"Lorem ipsum dolor sit amet, consectetur adipisicing elit, sed do eiusmod tempor incididunt ut labore et dolore magna aliqua. Ut enim ad minim veniam, quis nostrud exercitation ullamco

### GENERAL INFORMATION:

"Lorem ipsum dolor sit amet, consectetur adipisicing elit, sed do eiusmod tempor incididunt ut labore et dolore magna aliqua. Ut enim ad minim veniam, quis nostrud exercitation ullamco

### PROCEDURES:

   I. "Lorem ipsum dolor sit amet, consectetur adipisicing elit, sed do eiusmod tempor incididunt ut labore et dolore magna aliqua. Ut enim ad minim veniam, quis nostrud exercitation ullamco

   II. "Lorem ipsum dolor sit amet, consectetur adipisicing elit, sed do eiusmod tempor incididunt ut labore et dolore magna aliqua. Ut enim ad minim veniam, quis nostrud exercitation ullamco

   III. "Lorem ipsum dolor sit amet, consectetur adipisicing elit, sed do eiusmod tempor incididunt ut labore et dolore magna aliqua. Ut enim ad minim veniam, quis nostrud exercitation ullamco

   IV. "Lorem ipsum dolor sit amet, consectetur adipisicing elit, sed do eiusmod tempor incididunt ut labore et dolore magna aliqua. Ut enim ad minim veniam, quis nostrud exercitation ullamco
      A. "Lorem ipsum dolor sit amet, consectetur adipisicing elit, sed do eiusmod tempor incididunt ut labore et dolore magna aliqua. Ut enim ad minim veniam, quis nostrud exercitation ullamco

**FIGURE 1**  Document Format.

B.   "Lorem ipsum dolor sit amet, consectetur adipisicing elit, sed do eiusmod tempor incididunt ut labore et dolore magna aliqua. Ut enim ad minim veniam, quis nostrud exercitation ullamco.

**REFERENCES:**

**FORMS:**

**NOTES/ATTACHMENTS:**

**ORIGINATION DATE:**   #ORIGDATE#

**APPROVALS:**

| NAME | TITLE | DATE |
|------|-------|------|
| #APPROVER# | #APPRTITLE# | #APPRDATE# |

**FIGURE 1**   continued.

# 2 Policies and Plans

**Policy and Procedures**

Policy#: #POLNUM#
Location: #LOC#
Originating Department: #ORIGDEPT#
Effective Date: #EFFDATE#
Expiration Date: #EXPDATE#

## TITLE: ABUSE AND DOMESTIC VIOLENCE

### POLICY STATEMENT:

It is the policy of #ORG# that alleged or suspected abuse, neglect, or exploitation of an infant, child, elderly or disabled patient and all patients is reported to appropriate authorities. Adult patients experiencing family or domestic violence are provided information regarding family shelters and given additional information as required by law.

#ORG# adopts the #STATE# and (#ORG#) Child Abuse Screening, Assessment, Documenting and Reporting Policy by this reference and complies with all provisions of #ORG# policy.

### INTENT AND SCOPE:

This policy is intended to assist #ORG# personnel in identifying interpersonal violence and exploitation of all children, elderly and disabled patients and all patients coming in contact with the #ORG# and to set forth the corresponding reporting requirements, if any, of #ORG# personnel to the appropriate authorities.

## DEFINITIONS:

I. Medical Professional. "Medical Professional" means a Licensed Physician, Dentist, Nurse Practitioner, Physician Assistant, Registered Nurse, CRNA, LVN/LPN, Podiatrist or Social Worker.

II. Sexually Transmitted Disease. "Sexually Transmitted Disease" means any disease that is transmitted by any sexual activity as described in §§XX.X of the #STATE# Penal Code, whether reportable or not.

III. Mitigating Factors. "Mitigating Factors" means facts/circumstances considered relevant to the decision whether there is cause to report suspected abuse, the presence of which may indicate against a suspicion of abuse and/or neglect. Certain Mitigating Factors are documented for monitoring purposes, as set forth herein.

### GENERAL INFORMATION:

I. Any person who believes that a child, a person 65 years or older, an adult with disabilities and all patients is being abused, neglected, or exploited is required to report the circumstances to the #STATE#.

II. The #STATE# offers a central location to report:
    A. child abuse and neglect;
    B. abuse, neglect or exploitation of elderly or adults with disabilities;
    C. abuse of children in licensed child-care facilities or treatment centers for the entire State of #STATE#; and
    D. OTHER—specific to the #STATE#

III. A person making a report is immune from civil or criminal liability provided they make the report in good faith, and the name of the person making the report is kept confidential.

IV. Any person who suspects abuse and does not report it is subject to disciplinary action, up to and including immediate termination, and can be held liable for a Class B misdemeanor.

V. Timeframes for investigating reports are based on severity of allegations, as set forth herein.

VI. Reporting suspected child abuse makes it possible for a family to get help.

**PROCEDURES:**

## SCREENING PROCEDURES

I. Overview

    A. All patients are screened for abuse.

        1. Children, Disabled, Elderly. Any nurse, nurse practitioner, CRNA, physician assistant, podiatrist, dentist, or physician who suspects abuse, neglect, or exploitation of children, the disabled or elderly patients performs the following:

            a. Assess the patient in private and support the patient by encouraging realistic discussion of the situation and of patient's immediate safety needs. The presence of a Social Worker may be helpful to assist in the facilitation of the interview.

            b. Document the assessment and findings, including the location and extent of physical injury as well as the patient's statement of how the injury occurred. Statements from others, if applicable, as to how injury occurred are also documented.

            c. If required, report to the #STATE# as noted in their regulations. Document the report was made in the patient's chart.

            d. Sexual abuse is also reported through the #ORG# policy and procedure.

        2. Adults. Any Medical Professional who treats a person for injuries that the Medical Professional has reason to believe that an adult patient has been the victim of family or domestic violence:

            a. Immediately provides the person with written information regarding the nearest family violence shelter center. Use interpretation services as needed to communicate this information; and

            b. Documents reasons for this belief in the person's medical record and the information provided by the patient, as required by law.

    B. #ORG# Personnel contact the Security/Police Department if, in the employee's judgment, the patient's safety or level of care delivered is threatened.

    C. The Medical Professional caring for the patient in #ORG# performs the following:

        1. Assists and provides guidance, as needed, to staff in the formal reporting of alleged or suspected abuse, neglect, or exploitation to the appropriate regulatory agencies as required by state and federal law; and

        2. Documents all social assessments and actions taken in the patient's record.

II. Specific Circumstances per #STATE# and #ORG#:

    A. Circumstances, which trigger the responsibility of #ORG# to determine if a report of abuse is required, include but are not limited to:

        1. Minors who have never been married and are postpartum, pregnant, or have a child; and

        2. Minors who have never been married and who request a pregnancy test.

    B. Circumstances, which may trigger the responsibility to determine if a report of abuse is required, include but are not limited to:

        1. A minor who has never been married and is seeking birth control; and

            a. #ORG# does not require the contractor/provider to report abuse, neglect, or sexual abuse based solely on a minor's request for birth control.

            b. #ORG# determines whether acts have occurred which constitute abuse, neglect, or sexual abuse based on all information available from the routine treatment of the minor including, without limitation, whether any Mitigating Factors apply.

        2. A minor who has never been married and self-reports that he or she has a sexually transmitted disease (STD).

            a. #ORG# does not require a contractor/provider to report abuse, neglect, or sexual abuse based solely on the statement of a minor that he or she has an STD.

      b.  If test results from a physician or someone working under a physician's orders confirms a diagnosis of an STD:

          i.  #ORG# reports or determines and documents the presence of Mitigating Factors.

         ii.  However, without a medically confirmed diagnosis of an STD, #ORG# determines whether acts have occurred which constitute abuse, neglect, or sexual abuse based on all the information available from the routine treatment of the minor including, without limitation, whether any Mitigating Factors apply.

## REPORTING GUIDELINES

I. Persons Responsible for making report. The individual who conducts the screening and has cause to suspect abuse has occurred is legally responsible for reporting. A joint report may be made with the supervisor.

II. Timeframes for Reporting.

    A.  Professionals. Professionals, as defined herein, are required to report, not later than the XXth hour after the hour the professional has cause to believe the child has been or may be abused as defined in §XX.X of the #STATE# Code or is the victim of the offense of indecency with a child and the professional has cause to believe the child has been abused as defined in §XX.XX of the Family Code. For purposes of this sub-section, "professional(s)" means an individual who is licensed or certified by the state or who is an employee of a facility licensed, certified or operated by the state and who, in the normal course of official duties or duties for which a license or certification is required, has direct contact with children. A professional may not delegate to or rely on another person to make the report.

    B.  Nonprofessionals. Nonprofessionals make a report *immediately* after the nonprofessional has cause to believe that the child's physical or mental health or welfare has been adversely affected by abuse.

III. When to Report.

    A.  A report is made regardless of whether the staff suspects or knows that a report may have previously been made.

    B.  If the identity of the minor is unknown (e.g., the minor is anonymously receiving testing for HIV or an STD within the #ORG#), no report is required.

    C.  Emancipated minors are not subject to required reporting under §X.XX

    D.  Divorced minors are not subject to required reporting under §X.XX

IV. Where to Report.

    A.  Reports of abuse or indecency with a child are made to:

        1.  #STATE# Department (#STATE# Abuse Hotline at 1-###-###-#### operated 24 hours a day, 7 seven days a week or by fax to 1-###-###-####

        2.  Any local or state law enforcement agency;

        3.  The state agency that operates, licenses, certifies, or registers the facility in which the alleged abuse or neglect occurred; or

        4.  The agency designated by the court to be responsible for the protection of children.

    B.  When the alleged or suspected abuse or neglect involves a person responsible for the care, custody, or welfare of the child, a report is made to protective services.

V. What to Report.

    A.  The following are reported:

        1.  Name and address of minor, if known;

        2.  Name and address of the minor's parent or person responsible for the care, custody or welfare of the child if not the parent, if known; and

        3.  Any other pertinent information concerning the alleged or suspected abuse, if known.

    B.  If #ORG# does not routinely collect other kinds of information Protective Services or local law enforcement may request, the #ORG# is not required by law or the #ORG# policy to ask the client for that information.

VI. Anonymous Reporting.

    A.  Reports can be made anonymously.

B.  Documentation for #ORG# abuse does not prohibit a person from reporting anonymously. Documentation is a part of #ORG#'s records and is subject to the same protection of confidentiality as other client records.

VII. Confidentiality.

A.  #ORG# may not reveal whether or not the child has been tested for or diagnosed with HIV or AIDS.

B.  Reporting requirements apply regardless of professional confidentiality and licensing laws and rules for professionals. It is not a breach of confidentiality to report child abuse.

C.  For those programs also governed by federal laws, regulations, and policies, the federal grantors do not consider it a breach of confidentiality to follow state laws on reporting of child abuse.

D.  §XX.XX provides that reports of abuse to Protective Services as well as the identity of the person making the report are confidential and may be disclosed only for purposes consistent with applicable state or federal law or regulations.

VIII. Training. #ORG# provides training to its employees, volunteers, and medical staff on reporting a victim of abuse who is an unmarried minor under fourteen (14) years of age and is pregnant or has a confirmed STD acquired in a manner other than through perinatal transmission or transfusion.

IX.  Other Reporting. The #STATE# Code, requires reporting of other instances of sexual abuse. Other types of reportable abuse include but are not limited to, the actions described in:

A.  #STATE# Penal Code §XX.XX, relating to indecency with a child

B.  #STATE# Penal Code §XX.XX defining "sexual contact,"

C.  #STATE# Penal Code §XX.XX, defining various sexual activities

D.  #STATE# Penal Code §XX.XX relating to sexual assault of a child

E.  #STATE# Penal Code §XX.XX relating to aggravated sexual assault of a child

## DOCUMENTING GUIDELINES

I.  Mitigating Factors. For #ORG# monitoring purposes, #ORG# documents the presence of certain Mitigating Factors discovered during the screening process:

A.  Minors under the Age of Seventeen (17). #ORG# documents whether a client:

1.  under the age of seventeen (17) is or has been married and is therefore not a minor under #STATE# law;

a.  The #ORG# may, but is not required to, request documentation that a minor is married. #ORG# may choose to rely on statements by the minor as to his/her marital status.

2.  who has never been married, is under the age of seventeen (17), and who was determined to have been abused or neglected as defined by the §XX.XX, including but not limited to, victims of an offense under the Penal Code §XX.XX.

a.  The #ORG# may, but is not required to, request documentation that a minor is married. The #ORG# may choose to rely on statements by the minor as to his/her marital status.

b.  **Minors under the Age of Fourteen (14).** There are no Mitigating Factors for abuse of a minor under the age of fourteen (14).

c.  Age Difference.

d.  Sexual Indecency. Mitigating Factors for abuse as defined in the Penal Code §XX.XX (sexual indecency with a child) may be that the person was not more than three years older than the victim; and of the opposite sex; and the person did not use duress, force or a threat against the victim at the time of the offense.

e.  Sexual Assault. Mitigating Factors for abuse as defined in the Penal Code §XX.XX (sexual assault) may be that the person was not more than three years older than the victim.

## TRAINING

I. #ORG# training program for all #ORG# Personnel includes policies and procedures in regard to assessment of all patients for abuse, and documenting and reporting all abuse. New staff receives this training as part of their initial training/orientation and is a part of staff annual training. #ORG# requires that all training sessions be documented.

II. As part of the training, staff is informed that the staff person who conducts screening/assessment and has cause to suspect abuse has occurred is legally responsible for reporting. A joint report may be made with the supervisor.

## AGENCY JURISDICTIONS AND HOTLINE NUMBERS

I. The #STATE# protects children, the elderly, and people with disabilities from abuse, neglect, and exploitation by involving clients, families and communities.
   A. Call the Abuse Hotline toll-free 24 hours a day, 7 days a week, nationwide 1-###-###-####, or
   B. Report through a secure web site www.####.com (e-mail reports are not accepted),
   C. Hearing Impaired or Speech Disabled—People who are deaf, hard-of-hearing, or speech-disabled may report by using a TTY and accessing Relay #STATE# by dialing 1-###-###-####.

II. #STATE# investigates allegations of abuse, neglect, and exploitation in facilities that care for adults including: private homes, adult foster homes (with 3 or fewer consumers), unlicensed room and board, state facilities and community centers that provide mental health and mental retardation services, home health agency staff, exploitation in nursing homes when the alleged perpetrator is someone outside the facility.

III. #STATE# Departments that link to Aging and Disability

INSERT CONTACT INFORMATION FOR YOUR LOCAL OFFICE

IV. #STATE# Department of Health (#ORG#).

INSERT CONTACT INFORMATION FOR YOUR LOCAL OFFICE

V. #STATE# Abuse Reporting

INSERT CONTACT INFORMATION FOR YOUR LOCAL OFFICE

**REFERENCES:**


**FORMS:**

Reference the abuse form that is used by all professionals for assessment, intervention, reporting and regulatory documentation.


**EQUIPMENT:**


**APPROVALS:**

| NAME | TITLE | DATE |
| --- | --- | --- |
| #APPROVER# | #APPRTITLE# | #APPRDATE# |

**Policy and Procedures**

Policy#: #POLNUM#
Location: #LOC#
Originating Department: #ORIGDEPT#
Effective Date: #EFFDATE#
Expiration Date: #EXPDATE#

## TITLE: ADVANCE DIRECTIVES

### POLICY STATEMENT:

It is the policy of #ORG# to honor all Advance Directives issued by a patient in accordance with the laws of the State of #STATE#. It is the policy of #ORG# to inform every patient of the organization's policies regarding Advance Directives and communicate to patients all procedures that the institution is unwilling or unable to provide or withhold as relates to an Advance Directive.

### INTENT AND SCOPE:

This policy is intended to inform patients of their rights regarding Advance Directives, to communicate #ORG#'s commitment to honoring Advance Directives and to notify patients of any and all limitations and procedures related to providing and/or withholding care as relates to Advance Directives. This policy applies to all Medical staff and employees (contract and non-contract) of the #ORG#.

### DEFINITIONS:

I. Adult—A person eighteen (18) years of age or older or a person under eighteen (18) years of age who has had the disabilities of minority removed.

II. Advance Directive—A Directive to Physicians, Durable Power of Attorney for Health Care, Medical Power of Attorney, Out of Hospital-Do Not Resuscitate Order, or Declaration for Mental Health Treatment.

III. Agent—An adult to whom authority to make health care decisions is delegated under a medical power of attorney.

IV. Artificial Nutrition and Hydration—The provision of nutrients or fluids by a tube inserted in a vein, under the skin in the subcutaneous tissues, or in the stomach (gastrointestinal tract).

V. Attending Physician—A physician selected by or assigned to a patient who has primary responsibility for a patient's treatment and care.

VI. Cardiopulmonary Resuscitation—Any medical intervention used to restore circulatory or respiratory function.

VII. Competent—Possessing the ability, based on reasonable medical judgment, to understand and appreciate the nature and consequences of a treatment decision, including the significant benefits and harms of and reasonable alternatives to a proposed treatment decision.

VIII. Declarant—A person who has executed or issued a directive

IX. Declaration for Mental Health Treatment—Lists instructions for consent to or refusal of mental health treatment.

X. Directive to Physicians—Directs the attending physician to administer, withhold, or withdraw life-sustaining treatment should a person be certified in writing by the attending physician as suffering from a terminal or irreversible process.

XI. Durable Power of Attorney for Health Care—Directs a designated third party, as one's agent for purposes of making any health care decision should the person making the directive become incompetent per #STATE# state law.

XII. Ethics Committee—The Ethics Committee established for #ORG#.

XIII. Health Care or Treatment Decision—Consent, refusal to consent, or withdrawal of consent to health care, treatment, service, or a procedure to maintain, diagnose, or treat an individual's physical or mental condition.

XIV. Health Care Provider—(1) a hospital (2) a surgery center or other ambulatory program (3) a home and community support services agency (4) a personal care facility (5) a physician or other licensed healthcare professional and (6) a special care facility.

XV. Healthcare Professional—physicians, physician assistants, CRNAs, nurse practitioners, nurses, dentists, podiatrists, and emergency medical service personnel.

XVI. Incompetent—Lacking the ability, based on reasonable medical judgment, to understand and appreciate the nature and consequences of a treatment decision, including the significant benefits and harms of and reasonable alternatives to the proposed treatment decision.

XVII. Irreversible Condition—A condition, injury, or illness as determined by the attending physician (a) that may be treated but is never cured or eliminated (b) that leaves a person unable to care for or make decisions for the person's own self; and (c) that, without life-sustaining treatment provided in accordance with the prevailing standard of medical care is fatal.

XVIII. Life-Sustaining Treatment—Treatment that, based on reasonable medical judgment, sustains the life of a patient and without which the patient will die. The term includes both life-sustaining medications, and artificial life support, including but not limited to mechanical ventilation, renal dialysis treatment, and artificial nutrition and hydration. The term does not include the administration of pain management medication or the performance of a medical procedure considered to be necessary to provide palliative care.

XIX. Medical Power of Attorney—Directs a designated third party, as one's agent for purposes of making any health care decision should the person making the directive become incompetent.

XX. Out of Hospital DNR Order—(A) a legally binding out-of-hospital do-not-resuscitate order, prepared and signed by the attending physician of a person, that documents the instructions of a person or the person's legally authorized representative and directs healthcare professionals acting in an out-of-hospital setting not to initiate or continue the following life sustaining treatment: (1) cardiopulmonary resuscitation; (2) advanced airway management; (3) artificial ventilation; (4) defibrillation; and (5) transcutaneous cardiac pacing; (B) does not include authorization to withhold medical interventions or therapies considered necessary to provide comfort care, to alleviate pain or to provide water or nutrition.

XXI. Out of Hospital Setting—A location in which healthcare professionals are called for assistance, including long-term care facilities, private homes, hospital outpatient or emergency departments, physician's offices, ambulatory care settings and vehicles during transport.

XXII. Physician—(1) A physician licensed by the #STATE# State Board of Medical Examiners; or (2) A properly credentialed physician who holds a commission in the uniformed services of the United States and who is serving on active duty in this state.

XXIII. Principal—An adult who has executed a medical power of attorney.

XXIV. Proxy—A person designated and authorized by a Directive to Physicians executed or issued to make a treatment decision for another person in the event the other person becomes incompetent or otherwise mentally or physically incapable of communication.

XXV. Qualified Relatives—Those persons authorized by a Directive to Physicians executed or issued to make a treatment decision for another person in the event the other person becomes incompetent or otherwise mentally or physically incapable of communication.

XXVI. Statewide Out of Hospital DNR Protocol—A set, statewide, standardized procedure for withholding cardiopulmonary resuscitation and certain other life-sustaining treatment by healthcare professionals acting in out-of-hospital settings.

XXVII. Terminal Condition—An incurable condition caused by injury, disease, or illness that according to reasonable medical judgment as determined by the attending physician will result in death within six (6) months, even with available life-sustaining treatment provided in accordance with the prevailing standard of medical care. A patient who has been admitted to a program under which the person receives hospice services provided by a home and community support services agency is presumed to have a terminal condition.

## GENERAL INFORMATION:

I. Advance Directives are addressed in all adults (eighteen (18) years and older) in the ambulatory care setting.

II. Advanced Directives are not addressed in patients under eighteen (18) years, unless emancipated.

III. Advance Directives are addressed in the outpatient setting in a patient specific, physician driven manner.

IV. The original of any advance directive is returned to the patient or the authorized representative and a copy is placed in the current Medical Record.

V. Witnessing Provisions for Advance Directive:
  A. Is a document voluntarily executed by the declarant and signed by the declarant in the presence of two witnesses, who also sign the document.
  B. The two (2) witnesses are **not**:
    1. Related to the declarant by blood or marriage or entitled to any part of the declarant's estate after the declarant's death under a will or codicil executed by the declarant or by the operation of law.
    2. The physician caring for the patient or any employee of the physician.
    3. An employee of a health care facility with which the declarant is a patient, if the employee is providing direct patient care to the declarant or is directly involved in the financial affairs of the facility.
    4. A patient in a health care facility in which the declarant is a patient
    5. Any person who, at the time the Directive is executed, has a claim against any part of the declarant's estate after the declarant's death.

VI. Witnessing Provisions for Advance Directives:
  A. Each Witness is a competent adult; AND
  B. At least **one** of the Witnesses is a person who is **not**:
    1. designated by the Declarant to make a treatment decision;
    2. related to the Declarant by blood or marriage;

3. entitled to any part of the Declarant's death under a will or codicil executed by the Declarant or by operation of law;

4. the attending physician;

5. an employee of the attending physician;

6. an employee of a health care facility in which the Declarant is a patient if the employee is providing direct patient care to the Declarant or is an officer, director, partner, or business office employee of the health care facility or of any parent organization of the health care facility; or

7. A person who, at the time the written Advance Directive is executed or, if the Directive is a nonwritten Directive issued under this chapter, at the time the nonwritten Directive is issued, has a claim against any part of the Declarant's estate after the Declarant's death.

VII. When a patient presents you with paper work regarding advance directives you make a copy and place in the patient's record and return the originals to the patient.

VIII. If the patient fills out the form, which indicates they have received their patient's rights and responsibility booklet, then document that in the patient's medical record.

## DOCUMENTING OF PATIENT RESPONSIBILITY/ADVANCE DIRECTIVE STATUS

I. Patient Rights/Directives are documented in the patient's medical record and is always completed prior to obtaining an informed consent.

II. Patient Rights/Directives documentation and education is initiated prior to the day of surgery in the outpatient setting and is documented as such.

III. Follow-up and further documentation is completed as required per the patient's plan for care as directed by the primary physician.

## PROCEDURE FOR COMPLETING AN ADVANCE DIRECTIVE

I. Directive to Physicians;
A. Written Directives;
1. A competent adult may at any time execute a written Directive.
2. The Declarant signs the Directive to Physicians in the presence of two Witnesses in the manner outlined by this policy. The Witnesses sign the Directive.
3. A specific form is not required for a Directive to Physicians to be valid. No notarization is required for a Directive to Physicians to be valid.
4. Directive to Physicians are honored if properly witnessed.
5. Directives may contain directions from the Declarant. Directive to Physicians and a Declarant may designate a person to make a treatment decision for the Declarant in the event the Declarant becomes incompetent or otherwise mentally or physically incapable of communication.
6. The Declarant notifies the attending physician of the existence of a Directive to Physicians. If the Declarant is incompetent or otherwise mentally or physically incapable of communication, another person may notify the attending physician of the existence of the Directive to Physicians.
7. The attending physician makes the Directive a part of the Declarant's medical record.
B. Non-written Directives by Competent Adult Qualified Patient.
1. A competent qualified patient who is an adult may issue a Directive to Physicians by non-written means of communication.
2. A Non-written Directive is made by a Declarant in the presence of the attending physician and two witnesses who qualify under this policy.
3. The attending physician makes the fact of the existence of the Directive part of the Declarant's medical record and records the names of the Witnesses in the medical record.

C. Directive to Physicians for Patients Younger than 18 Years of Age, if NOT emancipated. The following individuals may execute a Directive on behalf of a qualified patient who is younger than 18 years of age and not emancipated:
   1. The patient's spouse, if the spouse is an adult.
   2. The patient's parents
   3. The patient's legal guardian
D. Procedure When Declarant is Incompetent or Incapable of Communication—Directive Executed:
   1. If the adult qualified patient has designated a person to make a treatment decision under a Directive to Physicians, the attending physician and the designated person may make a treatment decision in accordance with the Declarant's directions.
   2. If the adult qualified patient has not designated a person to make a treatment decision, the attending physician complies with the Directive to Physicians unless the physician believes that the Directive does not reflect the patient's present desire or previously expressed wishes.
E. Procedure When Declarant is Incompetent or Incapable of Communication—No Directive Executed
   1. If an adult qualified patient has not executed a Directive to Physicians and is incompetent or otherwise mentally or physically incapable of communication, the attending physician and the patient's legal guardian or an agent under a medical power of attorney may make a treatment decision that may include a decision to withhold or withdraw life-sustaining treatment from the patient in appropriate clinical circumstances.
   2. If the patient does not have a legal guardian or an agent under a medical power of attorney, the attending physician and one person, if available, from one of the following categories, in the following priority, may make a treatment decision that may include a decision to withhold or withdraw life-sustaining treatment:
      a. the patient's spouse
      b. the patient's reasonably available adult children
      c. the patient's parents
      d. the patient's nearest living relative
   3. All decisions are based on the patient's previously expressed wishes, if known.
   4. A treatment decision made under (E)(2) of this section is documented in the patient's medical record and signed by the attending physician.
   5. If the patient has no legal guardian and a person listed in (E)(2) of this section is not available, a treatment decision made under (E)(2) is affirmed by another physician who is not involved in the treatment of the patient or who is a representative of an ethics or medical committee of health care facility in which the person is a patient.
F. The desires of a qualified patient, including a qualified patient younger than 18 years of age, supersede the effect of a Directive to Physicians.
G. Prerequisites for Complying with Directive to Physicians.
   1. An attending physician who has been notified of the existence of a Directive to Physicians provides for the Declarant's certification as a qualified patient with a terminal condition or irreversible process.
   2. Before withholding or withdrawing life-sustaining treatment from a qualified patient, the attending physician determines that the steps proposed are in accord with #STATE# law, this policy, and the patient's previously expressed wishes.
H. Revocation or Re-execution of Directives
   1. All Directives to Physicians are valid until they are revoked.
   2. A Declarant may revoke a Directive at any time without regard to the Declarant's mental state or competency.
   3. A Directive may be revoked in the following manner:
      a. The Declarant or someone in the Declarant's presence and at the Declarant's direction cancels or destroys the Directive;
      b. The Declarant signs and dates a written revocation that expresses the Declarant's intent to revoke the Directive; or
      c. The Declarant orally states the Declarant's intent to revoke the Directive.

  4. In order to be valid, a Revocation of a Directive is communicated to the attending physician. Upon notice of revocation, the attending physician writes 'VOID' on each page of the copy of the Directive in the patient's medical record.

  5. A Declarant may at any time re-execute a Directive in accordance with #STATE# law including re-execution after the Declarant is diagnosed with a terminal or irreversible process.

 I. Pregnant Patients:

**NOTE:** A person may not withdraw or withhold life-sustaining treatment under this policy from a pregnant patient.

 J. Refusal to Honor Directive to Physicians:

**NOTE:** If for any reason a physician or health care professional fails to honor a Qualified Patient" Directive to Physicians, that physician or health care professional initiates #ORG# Policy "Refusal to Honor Advance Directive"

 A. Medical Power of Attorney:

 B. Form, Scope and Duration of Authority for Agent;

  1. The form of Information Concerning the Medical Power of Attorney and the Medical Power of Attorney is substantially similar. In no case is a Medical Power of Attorney effective without a signed statement from the patient that they have received, read and understood the Information Concerning the Medical Power of Attorney.

  2. The Agent may make any healthcare decision on the principal's behalf that the principal could make if the principal were competent, subject to limitations on the Agent contained in the Medical Power of Attorney.

  3. An Agent may exercise authority under the Medical Power of Attorney only after the attending physician certifies in writing and files the certification in the patient's medical record that based on reasonable medical judgment; the patient is incompetent or lacks decision-making capacity.

  4. Treatment may not be given to or withheld from the patient if the patient objects. It does not matter whether or not a Medical Power of Attorney has been executed by the patient or whether or not the patient is competent.

  5. The attending physician makes all reasonable efforts to inform the patient of any proposed treatment, withholding of treatment or withdrawal of treatment before implementing a Medical Power of Attorney.

  6. Agents, after consulting with the attending physician and other health care providers, make a healthcare decision for the patient based upon the patient's previously expressed wishes, accounting for the patient's religious and moral beliefs with the decided best interest of the patient in mind.

  7. An Agent may never consent to:

   a. Voluntary Inpatient Mental Health Services;

   b. Convulsive treatment;

   c. Psychosurgery;

   d. Abortion; or

   e. Neglect of the patient through the omission of care intended to provide for the comfort of the patient.

  8. The Medical Power of Attorney is effective indefinitely unless it is revoked as discussed in this policy or until the patient becomes competent.

  9. If the Medical Power of Attorney has an expiration date and the patient is incompetent on that date, the medical Power of Attorney is effective until the patient becomes competent unless the Medical Power of Attorney is revoked as discussed in this policy.

 C. Persons Who May Not Exercise the Authority of an Agent. The following people may not exercise the authority of an Agent while the are serving as:

  1. the patient's healthcare provider;

  2. an employee of the patient's healthcare provider unless the person is a relative of the patient;

  3. the patient's residential care provider; or

  4. an employee of the patient's residential care provider unless the person is a relative of the patient.

 D. Execution of Medical Power of Attorney and Witnesses;

  1. Witness: The Medical Power of Attorney must be signed by the patient in the presence of two witnesses who qualify under the Witnesses section of this policy.

  2. Patient Unable to Sign: If a patient is physically unable to sign a Medical Power of Attorney, another person may sign at the patient's direct request and in the patient's presence. The same Witness requirements apply.

E. Revocation of Medical Power of Attorney
1. A Medical Power of Attorney may be revoked by:
   a. Oral or written notification at any time by the patient to the agent or a licensed or certified health or residential care provider or by any other act evidencing the patient's intent to revoke the power, regardless of whether the principal is competent.
   b. Execution by the patient of a subsequent Medical Power of Attorney; or
   c. The divorce of the patient and spouse, if the spouse is the patient's Agent, unless the Medical Power of Attorney provides otherwise.
2. A patient's licensed or certified healthcare provider who is informed of a patient's revocation of a Medical Power of Attorney immediately records the revocation in the patient's medical record and gives notice of the revocation to the Agent and any known healthcare providers currently responsible for the patient's care.

F. Disclosure of Medical Information subject to specific limitations in the Medical Power of Attorney, an Agent may, for the purpose of making a healthcare decision:
1. Request, review and receive any information, oral or written, regarding the patient's physical or mental health, including medical and hospital records;
   a. Execute a release or other document required to obtain the information; and
   b. Consent to the disclosure of the information.

G. Duty of Healthcare Provider
1. A patient's healthcare provider and an employee of the provider who knows of the existence of the patient's Medical Power of Attorney follow the directives of the patient's Agent to the extent that it is consistent with the desires of the patient and this policy.
2. A patient's healthcare provider who finds it impossible to follow the Agent's directives because of a conflict with this policy or the Medical Power of Attorney must inform the Agent as soon as is reasonably possible. The Agent may request another attending physician.
3. If for any reason a physician or healthcare professional refuses to honor a provision or withholding of life-sustaining treatment, that physician or healthcare professional initiates #ORG# Policy "Refusal to Honor Advance Directive."

H. Discrimination Not Permitted:

**NOTE:** #ORG# does not discriminate against any patient based upon their execution of or failure to execute a Medical Power of Attorney by charging a different rate to the patient, refusing to admit the person to the facility, or refusing to provide the patient with healthcare.

I. Liability for Healthcare Costs:

**NOTE:** A patient is liable for costs of healthcare decisions made by his Agent.

J. Guardianship Issues:
1. If a person has filed a Petition or Appointment of a Guardian for a patient, the Agent still has authority to make healthcare decisions for the patient, absent a court order.
2. If the patient has a Guardian who is different than the Agent named in the Medical Power of Attorney and the Guardian files a petition with the court to become the patient's Agent, the guardian has the sole authority to make health care decisions for the patient during the tendency of the court's proceedings, absent court order to the contrary.

II. Out of Hospital Do Not Resuscitate:
A. A competent person may execute a written Out-Of-Hospital DNR order directing healthcare professionals acting in an out-of-hospital setting to withhold cardiopulmonary resuscitation and certain other life-sustaining treatment designated by the Board.
B. The person signs the Out-Of-Hospital DNR in the presence of at least two witnesses that qualify under the Witnesses section of this policy.
C. Incompetent Patient with an Existing Directive to Physicians:
1. A physician may rely on an incompetent patient's previously issued Directive to Physicians as the person's instructions to issue an Out-Of-Hospital DNR order and places a copy of the Directive to Physicians in the patient's medical record. The physician signs the Out-Of-Hospital DNR order in

      lieu of the patient signature as required on the Out-Of-Hospital DNR form. If the patient's previously executed Directive to Physician names a proxy, the proxy may sign the Out-Of-Hospital DNR.

    2. If a patient is incompetent but previously executed or issued a Medical Power of Attorney designating an Agent, the Agent may make any decisions required of the designating person as to an Out-Of-Hospital DNR order and signs the order in lieu of the patient.

  D. Effective Date of Out-Of-Hospital DNR:

**NOTE:** An Out-Of-Hospital DNR is effective on its execution.

  E. Form of Out-Of-Hospital DNR:
1. A written Out-Of-Hospital DNR is required to be in a standard form specified by the Department of Health.
2. A photocopy or other complete facsimile of the original written Out-Of-Hospital DNR order is permitted and is honored as if it is an original.

  F. Nonwritten Out-Of-Hospital DNR
1. A competent person is an adult may issue a nonwritten Out-Of-Hospital DNR.
2. The patient issues the nonwritten Out-Of-Hospital DNR order in the presence of the attending physician and two witnesses who qualify under the Witnesses section of this policy.
3. The attending physician and witnesses sign the Out-Of-Hospital DNR order a part of the declarant's medical record and the names of the witnesses are entered in the medical record.

  G. Out-Of-Hospital DNR Orders on Behalf of Minors:

**NOTE:** The minor's parents, legal guardian or managing conservator may execute an Out-Of-Hospital DNR order on behalf of a minor.

  H. Desire of Patient Supersedes Out-Of-Hospital DNR Order:

**NOTE:** The desire of a competent patient, including a competent minor, supersedes the effect of an Out-Of-Hospital DNR order executed by or on behalf of the patient.

  I. Procedure when Patient is Incompetent or Incapable of Communication when patient has executed an Out-Of-Hospital DNR Order.
1. If an adult patient has designated a person to make a treatment decision under a Directive to Physicians, the designated person and the attending physician comply with the Out-Of-Hospital DNR order.
2. If an adult person has not designated a person to make treatment decision under a Directive to Physicians, the attending physician complies with the Out-Of-Hospital DNR order unless the physician believes that the order does not reflect the person's present desire.

  J. Procedure When Person Has Not Executed or Issued Out-Of-Hospital DNR Order and is Incompetent or Incapable of Communication.
1. If an adult person has not executed an Out-Of-Hospital DNR order and is incompetent or incapable of communication, the attending physician and the person's legal guardian, Proxy, or Agent having a Medical Power of Attorney may execute an Out-Of-Hospital DNR order on behalf of the person in appropriate circumstances.
2. If the person does not have a legal guardian, Proxy, or Agent under a Medical Power of Attorney, the attending physician and at least one qualified relative from a category listed under **Procedures I.E.2** of this policy, in order of priority, may execute an Out-Of-Hospital DNR order.
3. At least two witnesses, meeting the guidelines of the Witnesses section of this policy witness the execution of the Out-Of-Hospital DNR.
4. If there is no qualified relative available to act for the person under section I., F., 2 of this policy, an Out-Of-Hospital DNR order is affirmed by another physician who is not involved in the treatment of the patient or who is a representative of #ORG#'s Ethics Committee.

  K. Compliance With State and Local Out-Of-Hospital DNR Protocols
1. All healthcare professionals honor Out-Of-Hospital DNR orders in accordance with the statewide Out-Of-Hospital DNR protocols and locally adopted Out-Of-Hospital DNR the protocols not in conflict with the statewide protocol.
2. All other protocol requirements met by healthcare professionals involved in the care of the person executing the Out-Of-Hospital DNR.

L.  DNR Identification Device
1.  The presence of a DNR identification device in the form prescribed by the Board on the body of a person is conclusive evidence that the person has executed a valid out-of hospital DNR order.
2.  Health care professionals honor the DNR identification device as if a valid Out-Of-Hospital DNR order form were found in the possession of the person.
M.  Duration of Out-Of-Hospital DNR:

**NOTE:** An Out-Of-Hospital DNR is effective until it is revoked as described in this policy.

N.  Revocation of Out-Of-Hospital DNR Order:
1.  A person may revoke an Out-Of-Hospital DNR order at any time without regard to the person's mental state or competency in the following manners:
    a.  The person executing the Out-Of-Hospital DNR, or someone at the person's direction and in the person's presence, destroys the order form or removes the DNR identification device or the person executing the Out-Of-Hospital DNR order form orally revokes the order.
    b.  An individual who identifies himself or herself as the legal guardian, a qualified relative, or as the Agent under a Medical Power of Attorney of the person who executed the Out-Of-Hospital DNR order destroys the order form or orally revokes the Out-Of-Hospital DNR.
2.  An oral revocation by a legal guardian, qualified relative or Agent under a Medical Power of Attorney only takes effect if the legal guardian, qualified relative or Agent is at the scene.
3.  The attending physician or other healthcare professionals at the scene record the time, date, place and method of revocation of an Out-Of-Hospital DNR in the patient's medical record in accordance with statewide Out-Of-Hospital DNR protocol. The attending physician writes 'VOID' on each page of the copy of the order in the patient's medical record upon the order's revocation.
O.  Pregnant Persons:

**NOTE:** A person may not withhold cardiopulmonary resuscitation or certain other life-sustaining treatment from a person known by the healthcare professionals to be pregnant.


**PROCEDURES:**


**REFERENCES:**


**FORMS:**


**EQUIPMENT:**


**APPROVALS:**

| NAME | TITLE | DATE |
|---|---|---|
| #APPROVER# | #APPRTITLE# | #APPRDATE# |

**Policy and Procedures**

Policy#:  #POLNUM#
Location:  #LOC#
Originating Department:  #ORIGDEPT#
Effective Date:  #EFFDATE#
Expiration Date:  #EXPDATE#

## TITLE: COMPETENCY ASSESSMENT

### POLICY STATEMENT:

#ORG# establishes and maintains a program designed to document that individuals who render services are competent to provide such services.

### INTENT AND SCOPE:

This policy is intended to establish a reporting mechanism for competency documentation for all employees (contract and non-contract) of #ORG#.

### DEFINITIONS:

### GENERAL INFORMATION:

   I. Programs/documents to document competency may include but are not limited to:
   A.  Staff licensure/certification
   B.  Proctoring and/or Precepting
   C.  Written employee performance appraisals
   D.  Continuing education programs or job specific education or training
   E.  Skills assessment checklists
   F.  Ongoing quality monitoring activities
   G.  Written job descriptions
   H.  Unit and Organization Orientation checklists
   I.  Individual Education Profile

  II. #ORG# assesses each staff member's ability to meet the performance expectations stated in his or her job description. Competency assessment is on going. Assistance in compiling programs or documents can be obtained from the leaders of the organization, from Human Resource personnel and/or from education coordinators.

 III. Employee, upon hire, receives a copy of their job description. The document is reviewed and the employee acknowledges the expectations of the position by authenticating the job description. The employee's supervisor also signs that the employee received the document. An authenticated copy of the job description is placed in the employee's file.

## GUIDELINES FOR COMPETENCIES

   I. Competencies are:
   A.  Specific to the job title/description/appraisal tool
   B.  Specific to unit/department/patient population
   C.  In place for clinical and non-clinical staff

D.  Based on consideration of (patient assessment)
    1.  Degree of supervision
    2.  Complexity of patient condition and status
    3.  Skill requirement/technology used
    4.  Complexity of assessment required
    5.  Congruent with the expectations of #ORG#, its patients and customers.

Each employee receives a periodic review of new procedures, new techniques and new technology and equipment changes.

## PROCEDURES:

I.  Standard forms are recommended for use to document that new employees, transfers, cross functioning, contract/agency are assessed for competency in an on-going process within ninety (90) days after starting a new job. Competency is documented on the approved form and is filed in the employee file.

Annual assessment and documentation of employee competency are addressed at the same time as the performance appraisal.

## REFERENCES:

## FORMS:

## EQUIPMENT:

## APPROVALS:

| NAME | TITLE | DATE |
| --- | --- | --- |
| #APPROVER# | #APPRTITLE# | #APPRDATE# |

**Policy and Procedures**

Policy#:  #POLNUM#
Location:  #LOC#
Originating Department:  #ORIGDEPT#
Effective Date:  #EFFDATE#
Expiration Date:  #EXPDATE#

## TITLE: DISPOSITION OF SPECIMENS

### POLICY STATEMENT:

It is the policy of #ORG# that the disposition of specimens is performed in accordance with local and national practice and in compliance with regulatory body guidelines.

### INTENT AND SCOPE:

The policy is intended to communicate the processes for acceptable disposition of specimens in #ORG#. This policy applies to all Medical Staff and employees (contract or non-contract) of #ORG#.

### DEFINITIONS:

### GENERAL INFORMATION:

I. Unusable pharmaceuticals, including controlled substances, are those drugs that are outdated, contaminated, or have been in an isolation environment. Contaminated items are pharmaceuticals dispensed to a patient care area or an ancillary department and have been returned to the medication room partially used or in an opened container.

II. Outdated pharmaceuticals are sent to the return company to be processed for full or partial credit.

III. Outdated sterile water and normal saline are provided to applicable educational bodies for teaching purposes upon request of the instructor and discretion of the Pharmacy nurse.

### PROCEDURES:

I. When a medication becomes unusable, it is returned for processing or destruction by the pharmacy nurse.

II. When an item that is not a controlled substance becomes unusable, it is placed in the Expired Drug boxes located in the medication room.

III. The pharmacy nurse separates unusable pharmaceuticals into those that can be returned to the vendor for credit and those that must be destroyed.

IV. Items that are to be returned for credit are processed by the return company.

V. Items that are to be destroyed are placed in red bags and destroyed by the Environmental Services Department according to established guidelines.
Disposition of controlled substances is done in accordance with DEA and EPA regulations.

**REFERENCES:**

**FORMS:**

**EQUIPMENT:**

**APPROVALS:**

| NAME | TITLE | DATE |
|------|-------|------|
| #APPROVER# | #APPRTITLE# | #APPRDATE# |

**Policy and Procedures**

Policy#:  #POLNUM#
Location:  #LOC#
Originating Department:  #ORIGDEPT#
Effective Date:  #EFFDATE#
Expiration Date:  #EXPDATE#

## TITLE: DISTRIBUTION AND CONTROL OF CONTROLLED SUBSTANCES

### POLICY STATEMENT:

It is the policy of #ORG# that all departments prescribing, dispensing, delivering and administering controlled substances comply with all statutes, regulations and standards governing such acts.

### INTENT AND SCOPE:

This policy is intended to provide guidelines that meet the legal and regulatory requirements for the control, distribution, documentation and administration of controlled substances. It affects all employees (contract and non-contract) of #ORG#.

## DEFINITIONS:

### GENERAL INFORMATION:

I. Controlled substances drugs are stored in a securely double locked, substantially constructed cabinet or automated dispensing device in the patient care areas.

II. Controlled substance drugs, I, II, III, IV, and V are stored in the designated pharmacy/medication room in a securely locked, substantially constructed cabinet, automated dispensing device, or vault. However, the organization may disperse such controlled substances throughout the stock of non-controlled substances in such a manner as to obstruct the theft or diversion of the controlled substances.

III. Controlled substances may not be transferred from one patient care area to another as a patient is transferred, with the exception of the use of an intravenous patient controlled analgesia unit or a pump controlled intravenous drip unit.

IV. Appropriate procedures to verify security and control are established and maintained in all areas in which controlled substance drugs are kept or administered.

V. When a patient is scheduled for a special procedure outside the physical boundaries of a department and:
   A. Is scheduled to return to the unit after the procedure and,
   B. A controlled substance drug is required.
   C. Medication may accompany the patient during the procedure provided the appropriate guidelines have been developed and approved to verify security and control.

VI. Procedures for distribution and control of controlled substances are formulated and disseminated by the Pharmacist.

VII. Completed controlled substance administration records and related documents, invoices or packing slips for controlled substance purchases are stored for a period of two (2) years as required by the Drug Enforcement Agency (DEA) regulations.

VIII. Outpatient pharmacy maintains prescription records for a period of at least two (2) years.

IX. All forms used in the documentation of distribution and administration controlled substance drugs are completed in their entirety in accordance with legal and regulatory agency criteria.

**PROCEDURES:**

**REFERENCES:**

**FORMS:**

**EQUIPMENT:**

**APPROVALS:**

| NAME | TITLE | DATE |
|------|-------|------|
| #APPROVER# | #APPRTITLE# | #APPRDATE# |

**Policy and Procedures**

Policy#: #POLNUM#
Location: #LOC#
Originating Department: #ORIGDEPT#
Effective Date: #EFFDATE#
Expiration Date: #EXPDATE#

## TITLE: DISTRIBUTION AND CONTROL OF STOCK MEDICATIONS

### POLICY STATEMENT:

It is the policy of #ORG# that the organization limits stock of drugs in patient care areas to those drugs appropriate to the care of the patient.

### INTENT AND SCOPE:

This policy communicates a standard approach for the distribution and control of floor stock pharmaceuticals both in the traditional secured method and in conjunction with Automated Dispensing Devices. This policy applies to all staff (contract and non-contract) of #ORG# authorized by law to administer medications.

### DEFINITIONS:

### GENERAL INFORMATION:

I. Par levels are established by patterns of usage.

II. The addition of new items to floor stock is done through a formal electronic request to the Pharmacy nurse and then authorized by the Medical Executive Committee and added to the organizations formulary.

III. Nursing Units:
   A. Floor stock items are loaded into appropriate Automated Dispensing Devices or locked in secured medication cabinets.
   B. Medications located in Automated Dispensing Devices are refilled daily as required.
   C. Medications located in secured medication cabinets are restocked by nursing personnel

### PROCEDURES:

I. Clinical Services:
   A. Stock Replacement;
      1. Staff prints daily medication refill reports, which are used to replenish automated dispensing system floor stock inventory PAR levels.
      2. The automated dispensing console automatically prints medication stock outs at the time they occur. Designated staff refill floor stock medication stock outs throughout the day.
      3. A Registered Pharmacist checks the medication refill for accuracy as noted in the contractual agreement for these services.
      4. Staff does not fill quantities in excess of par levels without approval from their supervisor.
      5. For replenishment of Stock medications in secure cabinets, nursing staff request approved PAR level medication quantities on a Stock Requisition Form submitted to clinical staff that is deemed competent

to perform this duty. The designated staff order requested medications at least twice a week to replenish the stock.

6. The automated dispensing system issues a medication charge to the patient when a nurse removes a medication from an automated dispensing device.

B. Each patient care area is billed for monthly shortages and credited for overstocked drugs.

II. Outpatient Clinic (OPC):

A. Each Clinic transmits an inventory and floor stock order to the designated staff once weekly. Stat orders are reviewed and filled as soon as possible. Orders can include both billable and non-billable medications.

B. The designated staff reviews the transmitted order for accuracy then sends to wholesaler for medication delivery.

C. The vendor fills order with requested medications and bills accordingly. Orders are delivered via Courier.

D. Each time a dose of medication is administered a charge is dropped in the patient care system by the clinic staff.

E. Overstocked medications are returned.

F. Shortages are ordered on the next unit stock order.

G. Expired meds are returned. Credit may be issued for medications returned to the vendor with a greater than six (6) month expiration date, provided the following conditions are met:

1. Drug wholesaler accepts the medication, and

2. Medication usable at another setting prior to expiration date.

Par levels are reviewed and adjusted quarterly.

**REFERENCES:**

**FORMS:**

**EQUIPMENT:**

**APPROVALS:**

| NAME | TITLE | DATE |
|------|-------|------|
| #APPROVER# | #APPRTITLE# | #APPRDATE# |

**Policy and Procedures**

Policy#:  #POLNUM#
Location:  #LOC#
Originating Department:  #ORIGDEPT#
Effective Date:  #EFFDATE#
Expiration Date:  #EXPDATE#

## TITLE: DOCUMENTATION OF CONTROLLED SUBSTANCE

### POLICY STATEMENT:

It is the policy #ORG# that drugs to be administered are verified with the prescribing credentialed practitioner's orders and are properly prepared and administered.

### INTENT AND SCOPE:

This policy is intended to provide a process for documentation of administration of controlled substances. This policy affects all Medical Staff and employees (contract and non-contract) of #ORG#.

### DEFINITIONS:

1. AMDS: Automated Medication Dispensing System.
2. MAR: Medication Administration Record.
3. PCA: Pain Control Administration
4. Wastage: Any portion of a controlled substance drug that has been opened but is not administered to a patient and is therefore discarded in an irretrievable form.

### GENERAL INFORMATION:

I. The official forms for the documentation of administration are a medication administration record (MAR); either printed or electronic, that is appropriate for the patient care area, and the Controlled Substance Administration Record.

II. In case an error in documentation is made and the entry needs to be altered, the person making the entry draws a single line through the entry and makes the documentation on the next available line. No attempt is made to obliterate the original entry. White out is not to be used for changes of any kind in a medical record.

III. Documentation for drug administration indicates the dose in metric quantities as in milligrams (mg) or milliliters (ml). No doses administered are documented in terms of the dosage form such as ampules or vial, without the specified metric quantity.

IV. Documentation of waste of controlled substances required the signature of two nurses on the Proof-Of-Use sheets.

### PROCEDURES:

I. The administering person documents on the controlled substance record the following:
   A.  The name of the drug.
   B.  The first and last name of the patient and the patient's room number or location.

 C. The name of the ordering physician.

 D. The amount of the drug given.

 E. The amount of the drug wasted if any.

 F. The date and time of administration.

 G. The name (signature) of the person administering the drug.

 H. The signature of a licensed witness to the waste.

II. The administration of the PCA is documented as follows:

 A. The cartridge is signed out on the Proof-Of-Use sheet.

 B. The nurse documents on the MAR or Flow Sheet the amount of medication the patient has received. This information is also documented on the PCA documentation record for the Narcotic drip.

 C. The number of attempts that the patient makes to self-administer after the drug is documented on the PCA documentation record.

 D. When the PCA is discontinued, the nurse makes an entry onto the PCA documentation record documenting the amount of drug wasted and obtains a wastage witness signature by a licensed nurse. The discontinuation of therapy is documented in the nursing record.

 E. When a patient receiving PCA therapy transfers to another unit, the transferring nurse documents on the PCA documentation record the amount of drug remaining. The nurse on the receiving unit documents the amount of drug remaining on the PCA documentation record upon the arrival of the patient to the unit.

## PROCEDURE FOR WASTING CONTROLLED SUBSTANCES:

I. When it is necessary to waste any amount of a controlled substance drug, a licensed nurse is summoned to the immediate area.

II. The person accessing the controlled substance drug measures and wastes the amount to be wasted while being observed by the licensed witness.

III. Both parties sign the record in the area designated for wasting/witnessing, wasting of a controlled substance drug.

IV. If all or part of a controlled substance taken from the Automated Medication Delivery Systems is wasted, nurse documents by choosing "waste option" under "Procedure", which requires two (2) nurse passwords to waste a controlled substance.

V. Pre-packaged unit-dose liquid narcotic substances cannot be returned for re-dispensing; therefore, these medications are wasted with an appropriate witness.

## NURSING AUDITS for NARCOTICS in LOCKED BOX:

I. At the beginning and end of the clinical day, and at change of shifts, licensed nurses conduct a controlled substance audit by counting the inventory of each controlled substance drug on the unit and comparing the count to the balance shown on the controlled substance documents.

II. Discrepancies (i.e., wastage signatures not present, count is incorrect, names omitted) are corrected immediately and prior to any nurses leaving the patient care area.

III. The balance at each count is written at the bottom of each Controlled Substance Proof-Of-Use sheet in the area designated "Narcotic Locked Box Count" and is signed by the two (2) nurses conducting the audit.

IV. The controlled substance form is placed in the Pharmacy Book after the audit is completed and the Pharmacy Book is locked up at all times to prevent tampering.

V. If discrepancies are not rectified immediately, a nurse comes to the Pharmacy and completes a Controlled Substance Variance Report for all unexplained losses or unresolved discrepancies.

VI. The Controlled Substance Variance Report is reviewed by the nursing area supervisor and is reviewed within twentyfour (24) hours.

VII. The consulting pharmacist reviews and audits all reports.

## NURSING AUDITS FOR CONTROLLED SUBSTANCES STORED IN AN AUTOMATED MEDICATION DELIVERY SYSTEM:

I. When the physical count for a controlled substance does not match the count displayed on the AMDS screen, a discrepancy is recorded in the machine and a report is printed.

II. The nurse continues the transaction by correcting the actual count so that the patient can receive their medication without delay.

III. The discrepancy is resolved with an appropriate witness by the end of the shift on which it occurred.

IV. If discrepancies are not rectified immediately, a nurse completes a Controlled Substance Variance Report for all unexplained losses or unresolved discrepancies.

V. The Controlled Substance Variance Report is reviewed by the nursing area supervisor within twentyfour (24) hours.

VI. Variance forms are reconciled against the medication printout by the designated nurses and audited/reviewed by the consulting pharmacist on a weekly basis.

## PHARMACY AUDITS USING PROOF OF USE SHEETS:

I. For incomplete or inaccurate Proof of Use Narcotic Sheets, the Director of Nursing returns a copy of the Proof of Use sheets to the appropriate Supervisor for each corresponding area for proper completion.

II. Completed Proof of Use copies are returned to the Director of Nursing within fortyeight (48) hours of the Supervisor's receipt.

III. Upon receiving a variance report of an un-rectified narcotic discrepancy form from an area, the clinical supervisor is alerted of the variance.

IV. The supervisor has fortyeight (48) hours to investigate and resolve the discrepancy.

V. If there is no resolution from the supervisor within fortyeight (48) hours, the Director of Nursing is notified.

VI. The Director of Nursing investigates and takes appropriate action against personnel involved in Narcotic variance.

VII. The consulting pharmacist reviews and acts on all discrepancies.

## PHARMACY AUDITS USING AUTOMATED MEDICATION DELIVERY SYSTEM:

I. For Controlled Substance discrepancies in the AMDS, the designated nursing staff runs a daily report of Undocumented Discrepancies, which is returned to the Nursing Area Supervisor for review.

II. Supervisors or licensed nursing staff resolves the discrepancies, according to policy, in the AMDS unit.

III. If the discrepancy cannot be resolved within the AMDS, a nurse completes a Controlled Substance Variance Report and forwards to the Supervisor within fortyeight (48) hours.

IV. Upon receiving a variance report of an un-rectified narcotic discrepancy form from an area, the supervisor is acts on the variance.

V. The supervisor has fortyeight (48) hours to investigate and resolve the discrepancy and if the discrepancy is not resolved, the Director of Nursing is notified.

VI. The Director of Nursing investigates and takes appropriate action against personnel involved in Narcotic variance.

VII. The consulting pharmacist reviews and acts on all discrepancies.

**REFERENCES:**

**FORMS:**

**EQUIPMENT:**

**APPROVALS:**

| NAME | TITLE | DATE |
|---|---|---|
| #APPROVER# | #APPRTITLE# | #APPRDATE# |

**Policy and Procedures**

Policy#:  #POLNUM#
Location:  #LOC#
Originating Department:  #ORIGDEPT#
Effective Date:  #EFFDATE#
Expiration Date:  #EXPDATE#

## TITLE: EMERGENCY MANAGEMENT PLAN

### POLICY STATEMENT:

It is the policy of #ORG# to provide personnel a plan to reduce the risk from an emergency or disaster. The plan includes processes to evaluate risks that may adversely affect the life or health of patients, staff, and visitors while in the course of its mission of providing a safe, secure, and therapeutic environment.

The plan is designed to support patient safety and effective care by providing reliable information that allows #ORG# management and staff to evaluate key issues and opportunities for improvement of emergency management performance. Effective planning is intended to reduce the impact of emergencies on the quality of patient care and increases the facility's ability to continue the provision of necessary patient care. Taken into consideration: space, personnel, supplies, communication and other essential services.

### INTENT AND SCOPE:

This policy is intended to meet legal, regulatory, licensing and accreditation requirements in accordance with federal, state, local, and regulatory agencies. The scope of #ORG#'s Emergency Management Plan is to describe how #ORG# establishes and maintains a program to ensure effective response to emergencies or disasters affecting the environment of care. This policy applies to Medical Staff, Volunteers and employees (contract and non-contract) of #ORG#.

Processes are defined to provide effective response to external and internal disasters and emergencies. "ORG" provides effective response to these disasters and emergencies. Community resources that may be called upon for help include: hospital emergency systems, the department of health, local utility companies, relief organizations such as the Red Cross and/or FEMA, and emergency medical systems.

### OBJECTIVES:

1. Ensure adequate supplies are available such as water, gases, food, supplies, medications and resources.
2. Provide alternate communication systems should the primary communication become ineffective.
3. Provide alternate utility sources.
4. Ensure safety of all staff, patients, physicians, practitioners, visitors and contract workers.
5. Protect and maintain the physical plant.
6. Define roles and responsibilities for personnel, including the medical staff and volunteers.
7. Create and implement a training program for all personnel, including the medical staff.

### DEFINITIONS:

ESO—Emergency Safety Officer appointed by administration.
EP—Emergency Plan.
CC—Command Center.
ICS—Incident Command System.
HVA—Hazard Vulnerability Analysis

IC—Incident Commander
OEM—Office of Emergency Management for the City

## GENERAL INFORMATION:

I. #ORG# has developed an Emergency Plan (EP) for managing six critical areas of emergency management:
   A. Communication
   B. Resources and assets
   C. Safety and security
   D. Staff responsibilities
   E. Utilities management
   F. Patient clinical and support activities

The Quality and Safety Committee is responsible for developing, implementing, and monitoring the Emergency Management program for #ORG#.

## PROCEDURES:

I. Representation from leadership and the medical staff to #ORG# actively participate in emergency management planning through the Environment of Care. A call tree is available that displays who is to be notified, how and by whom. Personnel emergency assignments are defined at each unit. For example: Admitting personnel process admissions and discharges, staff the bed control center, and, cancel all elective admissions and procedures; and, Security services controls traffic and media.

II. #ORG#'s HVA is evaluated annually via administration. Results of this evaluation results in performance improvement and safety opportunities. The assessment assists with revision of this plan, which then contributes to the New Year's plan. The revised plan is approved by the Board of Managers prior to implementation.

III. The #ORG# has consulted with municipal emergency management planners for prioritizing disaster preparedness. In reviewing the #CITY# HVA the potential for a ###### disaster event(s) was most likely to occur.

IV. The #ORG# leaders meets with the local emergency response agencies i.e., #CITY# Office of Emergency Management (OEM), and communicates needs and vulnerabilities. Based on the feedback from the OEM, the plan may be modified and/or disaster drills set up to meet the needs of the organization. The results of the risk assessment provide needed information so that the leadership team and the Emergency Safety Officer can effectively ameliorate and mitigate current and future emergency efforts.

V. The ESO coordinates efforts to identify resources that may be used if an emergency occurs. All departments and services participate in emergency preparedness efforts.

VI. The #ORG#'s organizational and management framework used to execute in a disaster response is modeled after the Incident Command System (ICS) model. The ICS is a systematic model that perpetuates the use of policies, personnel, plans, and, resources so that effective responses to disasters and emergencies can occur. Flexibility, scalability and rapid response are all characteristics of the ICS model for "ORG".

VII. The #ORG# maintains an ongoing inventory of resources and assets that may be used during an emergency i.e., decontamination equipment inventory, medical and surgical supply inventories, fuel, food, utilities, water/hydration, and medicine. This inventory is checked at least monthly with immediate actions taken should the approved inventory be compromised. The inventory is reviewed annually when the EP is reviewed. Adjustments are made as required to meet the facility's needs for disaster and emergency situations identified.

VIII. The #ORG#'s disaster planning involves all levels of the organization. Incident action plans and procedures are developed in collaboration with the all leaders, physicians, healthcare practitioners, staff, volunteers and contract services.

IX. #ORG#'s EP describes procedures for initiation and response for a variety of emergency action plans in the most direct means possible.

X. The effect from loss or failure could impact or disrupt business practices and the ability to deliver care. Efforts to quickly re-establish business, resume critical support functions, and continue the provision of care, are responded to with contingency plans and interim measures as needed. See the Backup Communication System policy, the Backup Water Supply policy, the Emergency Utilities Plan and the Medical Staff Roles and Responsibilities Policy along with the department specific plans.

XI. The EP is initiated by the administrator, and in the absence of the administrator, the Chain of Command policy is followed. The initiation process in each emergency plan is completed by the most direct means possible, either through overhead announcement, the alarm system, pagers, personal phones, or through messengers if all systems are down.

XII. The EP identifies the director of nursing and/or nursing supervisor serves as the incipient Incident Commander (IC) until properly relieved by the administrator. The Incident Commander has the authority to activate the response and recovery phases of the plan.

XIII. Alternate care and treatment sites to meet the needs of patients during emergencies are identified for care. These sites are ###############.

XIV. During a disaster all information and communications are funneled through the department leaders to the IC then disseminated back to these leaders for communicating to staff.

XV. The #ORG# works with representatives of community emergency response agencies in developing the OEM plan. A chart that identifies each community response agency, the critical emergency response activity, and agency roles are attached. Agencies are notified by the IC as soon as possible after an emergency response is initiated.

XVI. Normal telephone communication channels are first to be sought. If normal channels are not operative #ORG# uses all available means to communicate with agencies including employment of hand-held radio, runner, amateur radio, etc.

XVII. If #ORG# can no longer sustain operations and relocation of patients becomes necessary; #ORG# notifies family members of transfer and provides the name of the facility, and the name and telephone number of contact individuals at the facilities used for all transfers.

XVIII. When the EP is initiated #ORG# communicates with external agencies and media through the administrator or designee for any disaster event.

XIX. When emergency responses are initiated, #ORG# may utilize its vendors list for essential supplies, services and equipment as necessary. Vendors are notified by telephone (or other means if the telephone system is not operational) to respond to the organization's needs should they arise.

XX. #ORG# communicates with other health care organizations and provides the names, roles, and telephone numbers of our command structure for emergency management.

XXI. The #ORG# assesses resources and assets that can be shared with other local organizations in the event of a disaster. The decision to transfer resources and assets is made by the IC.

XXII. All patients are entered into a patient triage roster that is located in the triage area and continually updated and submitted to the Incident Commander and other health care organizations.

XXIII. The communicating of patient information with community third parties are performed in disaster events and in compliance with laws and regulations. Communication is coordinated through a collaborative effort between the CC and alternative care sites. The #ORG#'s alternative communication systems include runners, hand-held portable radios, amateur radios, landline phone, cellular phone, email, etc. Communication systems are imperative to the success of any exercise or actual event. Communication systems are routinely tested to ensure optimum performance.

XXIV. #ORG# continually monitors inventories for an extended emergency and during the recovery phase. The #ORG# through its Purchasing process and pharmaceutical relationships has suppliers replenish medical supplies and equipment. The #ORG# maintains supplies and medications that may be required for an extended emergency at all times. Administration may contact suppliers upon the onset of emergency and communicate the need for emergency replenishment of supplies.

XXV. Arrangements are in place so that patients may be transferred to a facility that can provide adequate continuum of care. Ambulances for the transfer of patients between facilities, licensed transportation services as well as other transportations services are identified in the calling tree.

XXVI. When the environment cannot support care, treatment and services, and the CC has ordered evacuation of the facility to an alternate care site, it will be necessary to transfer equipment, medications, essential clinical and medication-related information, and supplies to the alternate care site. This shall be coordinated through the CC and the transfer of these components is made utilizing transportation arrangements.

XXVII. Safety and security activities when emergency measures are initiated play a vital role during response and recovery phases of emergencies. Safety and security duties are assigned and activities for protection of patients, staff, and assets when emergency measures are initiated.

XXVIII. Command of security inside the building is under the IC. Community security agencies work collaboratively with internal security personnel.

XXIX. #ORG# has a temporary secured area for hazardous materials and waste storage until the emergency conditions have been lifted and the vendors are contacted to remove materials.

XXX. During lockdown conditions, #ORG# secures all entrances with the exception of the Main entrance. Only persons with proper identification will be admitted to the facility during an emergency.

XXXI. During emergency conditions, movement within the facility is controlled by assigned staff serving as security through security check points, control of elevators, control of doors, and controls to ensure that vital staff and equipment receive preference for reaching their intended destination.

XXXII. #ORG# accomplishes advanced preparation through mandatory staff education and disaster/emergency drills.

XXXIII. To provide safe and effective patient care during an emergency, staff roles are well defined in advance and assigned responsibilities with the use of job action sheets. These job sheets define essential staff functions.

XXXIV. The CC identifies areas and personnel to assist in family support needs. The Personnel Staging Area is contacted to determine availability of staff members to be assigned to staff these areas.

XXXV. #ORG# personnel and practitioners/physicians are identified during emergencies by their employee/practitioner badge. If an employee/practitioner does not have a badge, temporary badges can be obtained through Human Resource, upon verification. Additionally, extra staff/practitioners are sent to the IC to help the organization respond to the needs of the disaster. The IC assigns roles as requests are received. In the event volunteers from the community come to the facility, their credentials are verified.

XXXVI. #ORG# implements plans as the situation dictates. If there is advanced warning the Incident Commander implements the plan in advance to notify and possibly recall staff/practitioners as needed.

XXXVII. Alternative means of meeting essential utility needs such as power, medical gases, water, ventilation, and fuel are coordinated through Plant Operations/administration as part of the Utility Systems Management Plan. Additionally, external pre-arrangements for essential services are initiated. The facility has a fuel powered emergency generation system capable of providing for electricity in an emergency.

XXXVIII. Water needed for consumption and essential care activities has been calculated based on need and is stored on the facility's premises. During emergencies the facility implements a conservation policy.

XXXIX. Water needed for dialysis equipment, dish washing, instrument washing, hand washing and for other equipment and sanitary purposes, has been calculated based on need for an extended period and is stored on the premises. During emergencies #ORG# implements a conservation policy.

XL. Diesel fuel for operation of the emergency generators has been calculated and is stored on site to support an extended response period.

XLI. The piped medical gas systems are connected to emergency power. Provisions have been made for portable air cylinders and appropriate regulators to enable stand-alone operations where medical air may be required. The medical vacuum system is connected to an emergency power source and as a backup "critical" patient units are provided with battery operated suction pumps located on crash carts.

XLII. #ORG# has an emergency code for evacuation of the facility or unit. In the event a unit is deemed unsuitable for continued occupancy the code "Emergency Evacuation from *location*" is initiated. Reference the EP Evacuation Plan for details. Evacuation of the facility is performed in four parts: Visitors, Ambulatory Patients, Non-Ambulatory Patients, and Staff.

XLIII. #ORG# has plans in place to mitigate utility outages, i.e., bottled water for sanitation and hydration, radios and cell phones in place of landline phones, and alternate surgical sterilization facilities.

XLIV. Elective admissions and procedures are canceled, including elective surgery and non-emergency outpatient procedures. Stable patients are discharged.

XLV. Clinical activities for vulnerable patient populations (pediatric, geriatric, disabled, psychiatric and addiction) are provided in the customary way but additional emphasis is placed on security, safety, and mobility in terms of evacuation is in place should it become necessary during an emergency.

XLVI. Personal hygiene and sanitary needs of patients during emergencies is provided with available water.

XLVII. #ORG# collaborates with the OEM and #ORG# Medical Examiners Office management of services.

XLVIII. #ORG# is equipped with back up data systems designed to be retrieved during emergencies and utilized for documenting and tracking patients' clinical information. Additionally, paper forms are also utilized to document and track patients' clinical information if needed.

XLIX. Components of the EP are implemented as soon as notification of an emergency exists. Two emergency/disaster drills are completed yearly.

L. #ORG# grants disaster privileges to volunteer licensed independent practitioners only when the EP has been activated. The Medical Director and administrator may grant temporary privileges.

LI. Volunteers report to the Human Resources Department for issuance of temporary badges to identify them as volunteer independent practitioners.

LII. Licensed independent practitioners are directly under the supervision of the medical director or designee, either by direct observation or the assignment of a mentor from other staff physicians.

LIII. Primary source verification of licensure may not required if the volunteer licensed independent practitioner has not provided care, treatment, or services under the disaster privileges.

LIV. During disaster situations, individuals with specific medical licenses and skills that can be valuable to patient care of the community may report to the facility to provide volunteer assistance. These volunteers are directed to the IC to verify licensure. If licensure can be verified, the volunteers are used as necessary. If licensure cannot be verified, the volunteers can be used in roles that are not directly related to patient care.

LV. Volunteer Services in conjunction with Nursing Services is responsible for assigning disaster responsibilities to practitioners who are not licensed independent practitioners on a supply and demand basis. Volunteers reports to Human Resources Department for issuance of a badge identifying them as unlicensed volunteer practitioners.

LVI. Any unlicensed independent volunteer practitioner is directly under the supervision of the department staff, either by direct observation, assigning a mentor from other staff and/or medical record review or a combination of any of them.

LVII. During disaster situations, individuals with specific medical licenses and skills that can be valuable to patient care of the community may report to the facility to provide volunteer assistance. These volunteers will be directed to the Personnel Staging Area to verify licensure. If licensure can be verified, the volunteers are used as necessary. If licensure cannot be verified, the volunteers can be used in roles that are not directly related to patient care.

LVIII. Each volunteer is under the supervision of on-duty staff.

LIX. Primary source verification of licensure, certification, or registration is not required if the volunteer practitioner has not provided care, treatment, or services under his or her assigned disaster responsibilities.

LX. Each disaster/management exercise incorporates the most likely scenarios identified in #ORG#'s HVA; and, the components of the exercise are monitored and assessed, including communications, resources, assets, inventories, security, utilities and patients.

LXI. An individual who is not a participant in the exercise and whose sole responsibility is to monitor execution of the drill is assigned to each exercise. This individual is knowledgeable in the goals and expectations of the exercises and objectively documents opportunities for improvements.

LXII. Exercises are designed to test the EP and are as realistic as possible. Performance is monitored for effectiveness of staff notification of events, internal and external communications, availability and mobilization of resources, and effective and timely patient management.

LXIII. Each exercise is critiqued and opportunities for improvement are documented. Critique data from the exercises, are reviewed through members representing safety, administration, clinical services, physicians and support staff. The strengths and weaknesses identified in the exercise critiques are used to modify the Emergency Plan. The EP is reviewed and modified to reflect the necessary measures needed.

**REFERENCES:**


**FORMS:**


**EQUIPMENT:**


**APPROVALS:**

| NAME | TITLE | DATE |
|---|---|---|
| #APPROVER# | #APPRTITLE# | #APPRDATE# |

**Policy and Procedures**

Policy#: #POLNUM#

Location: #LOC#

Originating Department: #ORIGDEPT#

Effective Date: #EFFDATE#

Expiration Date: #EXPDATE#

## TITLE: EXPOSURE CONTROL PLAN FOR BLOODBORNE PATHOGENS

### POLICY STATEMENT:

It is the Policy of #ORG# to comply with federal guidelines pertaining to exposure control measures for Bloodborne Pathogens.

### INTENT AND SCOPE:

The intent of this policy is to provide guidelines that facilitate compliance with Occupational Safety and Health Administration (OSHA) standards for the prevention of occupational exposure to bloodborne pathogens.

## DEFINITIONS:

I. Bloodborne pathogens: Pathogenic microorganisms that are present in human blood and can cause disease in humans. These pathogens include, but are not limited to, Hepatitis B virus (HBV), Hepatitis C (HCV) and the Human Immunodeficiency Virus (HIV).

II. Standard Precautions: The use of barrier precautions when in contact with all blood and body fluids (except sweat) from all patients.

III. Contaminated: The presence, or the reasonably anticipated presence, of blood or other potentially infectious materials on an item or surface.

IV. Contaminated Laundry: Any laundry that may contain blood and/or other potentially infectious materials.

V. Contaminated Sharps: Any contaminated object that can penetrate the skin including, but not limited to, needles, scalpels, broken glass, broken capillary tubes, and exposed ends of dental wires.

VI. Decontaminated: The use of physical or chemical means to remove, inactivate, or destroy bloodborne pathogens on a surface or item to the point where they are no longer capable of transmitting infectious particles and the surface or item is rendered safe for handling, use or disposal.

VII. Engineering Controls: Controls (e.g., sharps disposal containers, self-sheathing needles, safer medical devices, such as sharps with engineered sharps injury protections and needleless systems) that isolate or remove the bloodborne pathogens hazard from the workplace.

VIII. Needleless Systems: A device that does not use needles for:
   A. The collection of bodily fluids or withdrawal of body fluids after initial venous or arterial access is established;
   B. The administration of medication or fluids; or
   C. Any other procedure involving the potential for occupational exposure to bloodborne pathogens due to percutaneous injuries from a contaminated sharp.

IX. Sharps with Engineered Sharps Injury Protection: A non-needle sharp or a needle device used for withdrawing body fluids, accessing a vein or artery, or administering medications, or other fluids, with a built-in safety feature or mechanism that effectively reduces the risk of an exposure incident.

X. Sharps Injury Log: establish and maintain a sharps injury for the recording of percutaneous injuries from contaminated sharps. The information in the sharps injury log is recorded and maintained in such manner as to protect the confidentiality of the injured employee. The sharps log contain, at a minimum;
   A. The type and brand of device involved in the incident,
   B. The department or work area where the exposure incident occurred;
   C. An explanation of how the incident occurred.

XI. Exposure Incident: An exposure incident is defined as a specific eye, mouth, or other mucous membrane, non-intact skin, or parenteral contact with blood and/or other potentially infectious material.

XII. Healthcare Worker: Healthcare Workers refers to all the paid and unpaid persons working in health care settings. This may include, but is not limited to, physicians, nurses, aides, dental workers, technicians, workers in laboratories and morgues, emergency medical service (EMS) personnel, students, part-time personnel, temporary staff not employed by the health-care facility, and persons not involved directly in patient care but who are potentially at risk for occupational exposure (i.e., volunteer workers, dietary, housekeeping, maintenance, clerical and janitorial staff).

XIII. Occupational Exposure: Reasonably anticipated skin, eye, mucous membrane, non-intact skin, or parenteral contact with blood and other potentially infectious materials that may result from the performance of an employee's duties.

XIV. Parenteral: Piercing mucous membranes or the skin barrier through such events as needle sticks, human bites, cuts and/or abrasions.

XV. Personal Protective Equipment (PPE): "Barriers include gloves, gowns, face shields, masks, protective eyewear, and ventilation devices."

XVI. Other Potentially Infectious Material (OPIM):
   A. Includes blood and human body fluids including "semen, vaginal secretions, cerebrospinal fluid, synovial fluid, pleural fluid, pericardial fluid, peritoneal fluid, amniotic fluid, saliva in dental procedures, any body fluid that is visibly contaminated with blood, and all bloody fluids in situations where it is difficult or impossible to differentiate between body fluids.
   B. Any unfixed tissue or organ (other than intact skin) from a human (living or dead).

XVII. Regulated Waste: Liquid or semi-liquid blood or other potentially infectious materials; contaminated items that would release blood and/or other potentially infectious materials in a liquid or semi-liquid state if compressed; items that are caked with dried blood or other potentially infectious materials and are capable of releasing these materials during handling; contaminated sharps; and pathological and microbiological wastes containing blood and/or other potentially infectious materials.

XVIII. Source individual: Any individual, living or dead, whose blood or other potentially infectious materials are a source of occupational exposure to the employee.

XIX. Work Practice Controls: Controls that reduce the likelihood of exposure by altering the manner in which a task is performed (e.g., prohibiting recapping needles by a two-handed technique.)

**GENERAL INFORMATION:**

I. Exposure Determination—The Clinical Leaders in conjunction with Infection Prevention Professionals compiles and maintains a list of job classifications of all employees who have a reasonable likelihood of occupational exposure. The list is kept in Human Resources.

II. Engineering Controls—Engineering controls are used to protect workers from exposure to bloodborne pathogens. These controls include, but are not limited to:
   A. Safety design devices
   B. Needleless systems
   C. Devices with engineered sharps injury protection
   D. Sharps containers
   E. Hand washing facilities
   F. Environmental Spills

III. The use of Needleless Systems, Needle Devices, Non-Needle Sharps represents a very efficient means of reducing potential injuries. Needleless system(s)/devices with engineered sharps protection are to be used for:
   A. Withdrawing OPIM after initial venous or arterial access is established,
   B. Accessing a vein or artery
   C. Administering fluids or medications and,
   D. Any other procedure involving the potential for an exposure incident for which a Needleless system is available as an alternative to using a needle device.
   E. Nursing personnel clean up any Bloodborne and/or OPIM spill on patients' beds.
   F. The center staff is responsible for cleaning up after any Bloodborne and/or OPIM spill in surgical and procedural suites.
   G. Staff sanitizes the area immediately after nursing personnel completes initial clean up.

IV. Engineered sharps injury protection devices are NOT used for the following reason(s):
   A. The direct patient caregivers who have participated in the trial studies of products have determined that the use of the engineering control would jeopardize patient or employee safety or may be unduly burdensome. In such cases, a committee assigned to this review documents this exception in their minutes and a request for a waiver is submitted to the #STATE# Department of Health Services if appropriate.
   B. The Committee evaluates various engineering controls and determine with evaluations from direct patient caregivers which products provide the best protection without compromising patient care. Devices selected by the evaluation committee are adopted.
   C. Employees with potential occupational exposure to blood and OPIM are trained in the use of engineering controls provided for their use. Additional training is provided as necessary when new engineering controls are adopted.
   D. New devices are evaluated annually and as appropriate.

V. Work Control Practices:
   A. The use of standard precautions is an integral part of the exposure control plan and of #ORG#'s Infection Control program. Standard Precautions are practiced.
   B. Work practice controls/procedures have been implemented to minimize exposure to bloodborne pathogens. Each department manager/supervisor is responsible for implementing, evaluating and monitoring compliance with these work practices by way of annual performance evaluations and competency checklists. Infection Control monitors work practices during rounds performed in each area.
   C. Follow-up on the report of an employee's failure to comply with the required protective measures is the responsibility of the employee's supervisor or director.
   D. Follow-up of problems identified through any means is the informal reports, complaints from staff, quality improvement or safety reports, minutes from committees, employee questionnaires, staff logs, and comments are the responsibility of the affected department's Director.
   E. Supervisors and employees examine and maintain work practice controls on a regular basis.
      1. The use of work practice controls are re-evaluated annually during the yearly review of this plan in conjunction with reported occupational exposure/injury data.

  2. Additions or deletions to work practice controls are made at that time or more often as indicated by ongoing monitoring activities.
  3. The following work practice controls are followed:
      a. Hand washing facilities are readily available in all work areas.
          i. If Hand washing facilities are not available when exposure to blood or OPIM occurs, employees may use an antiseptic cleanser in conjunction with clean paper/cloth towels, antiseptic towelettes or waterless disinfectant for immediate cleaning.
          ii. Hands are washed with soap and running water as soon as feasible. Refer to the #ORG# policy for Hand Hygiene
      b. Mucous membranes and eyes are washed with soap and water or flushed with water as appropriate as soon as feasible following exposure to blood or OPIM.
      c. #ORG# issued scrubs are to be covered with either a lab coat or jacket when leaving the designated area of use.
      d. Surgical instruments are handled through a safe zone.
F. Work Area Guidelines:
  1. Eating or drinking is permitted only in designated areas separate from contaminated areas.
  2. Food and/or drink are not kept in refrigerators, freezers, shelves, cabinets, or on countertops or bench tops where blood and/or other potentially infections materials are present.
  3. Applying cosmetics or lip balm and/or handling contact lenses is prohibited in work areas.
  4. Mouth piping/suctioning of blood or body substances is prohibited.
  5. All procedures are conducted in a manner that minimize splashing, spraying, splattering and generation of droplets of blood or OPIM.
G. Handling of Contaminated Sharps is outlined in Isolation Precautions #ORG# Policy
H. Handling of specimens:
  1. Specimens of blood, body fluid, and tissue are placed in biohazard specimen bags (secondary containers) to prevent leakage during collection, handling, processing, storage, transport or shipping. Place the requisition slip in the outside pocket.
  2. All specimens are handled as potentially infectious material and are transported consistent with the Laboratory Services Infection Control Policy.
I. Personal protective equipment (PPE) is an essential component of the plan to protect employees from exposure to bloodborne pathogens. Proper use of PPE is a condition of employment.
  1. Personal protective equipment is used in conjunction with, not as a substitute for, engineering controls and work practice controls.
  2. Where potential occupational exposure exists, PPE is provided. Examples of PPE include gloves, eyewear with side shields, gowns, lab coats, aprons, shoe covers, face shields and masks.
  3. PPE is selected and worn based on anticipated exposure to blood or OPIM.
  4. Appropriate PPE is fluid resistant under normal conditions and use time and does not permit blood or OPIM to pass through or reach the employee's clothing, skin, eyes, mouth or other mucous membranes.
  5. #ORG# is responsible for cleaning, laundering, repairing, replacing and discarding PPE. Employees are not allowed to launder PPE.
  6. All garments, which are penetrated by blood, body fluids or OPIM, are removed immediately or as soon as feasible and placed in the appropriated container.
  7. ALL PPE IS REMOVED PRIOR TO LEAVING THE WORK AREA.
  8. Employees are responsible for placing PPE in the proper designated receptacle for storage, washing, decontamination or disposal.
  9. Employees wear gloves when it is reasonably anticipated that they have hand contact with blood or OPIM, non-intact skin or mucous membranes. Latex-sensitive employees are provided with suitable alternatives such as latex-safe gloves, glove liners and powder less gloves.
  10. Disposable gloves are not washed or decontaminated for future use.
  11. Heavy-duty utility gloves may be decontaminated for re-use. Utility gloves are discarded if they are cracked, peeling, torn, punctured, exhibit any signs of deterioration or can no longer function as a barrier.
  12. Employees wear masks in combination with eye protection such as goggles, glasses with solid side shields or chin-length face shields whenever splash, spray, splatter or droplets of blood or OPIM may be generated, and eye, nose or mouth contamination reasonably can be anticipated.

VI. Housekeeping
   A. The work-site is maintained in a clean and sanitary condition. An appropriately written schedule for cleaning and method of decontamination is maintained based on the location, type of surface to be cleaned, type of soil and tasks or procedures performed in the area. Refer to the Environmental/Housekeeping Manual and the individual departmental policy/procedure manuals.
   B. All contaminated work surfaces are decontaminated:
      1. After completion of each procedure except in situations where procedures are performed on a continual basis throughout the shift such as blood analyses.
      2. When they are obviously contaminated during a procedure.
      3. Immediately or as soon as possible when surfaces are obviously contaminated.
      4. After any spill of blood or other potentially infectious materials, Environmental Services is responsible for clean up.
      5. At the end of the work shift if the surface may have been contaminated since the beginning of the shift.
   C. Cleaning supplies are available for employees to use when housekeeping staff is not available. Employees are trained within their department in the appropriate procedure for cleaning up blood spills.
   D. Trash containers with plastic liners are used for contaminated items and are routinely inspected and decontaminated on a regular schedule or when visibly contaminated.
   E. Broken glassware is not picked up directly with hands. It is cleaned up using mechanical means such as a brush and dustpan, tongs, or forceps.

VII. Laundry:
   A. All used and soiled linen is considered potentially infectious and is handled as contaminated and in accordance with Isolation Precautions Policy.
   B. Refer to Textile Processing Policies as listed in Reference section for specific information of sorting, handling and transporting procedures.

VIII. Waste Disposal:
   A. Employees are trained to close containers prior to removal to prevent spillage during handling, transporting or shipping.
   B. Refer to #ORG# Policy on Waste Disposal.

IX. Compliance Monitoring:
   A. The Safety Committee reviews all exposure incidents to determine the root cause of incidents and makes recommendations as appropriate to prevent the occurrence of similar incidents in the future.
   B. Recommendations are reported to the Quality Committee along with the affected department.
   C. The affected department is responsible for implementing corrective measures.
   D. Noncompliance Evaluation:
      1. Noncompliance is reported to an employee's immediate supervisor.
      2. The employee's immediate supervisor documents noncompliance.
      3. The effectiveness of corrective actions, such as providing more training for a non-compliant employee or identifying and correcting a deficiency with engineering controls, is documented.

X. Hazard Communication:
   A. Warning labels are affixed to containers of regulated waste, refrigerators and freezers containing blood or other potentially infectious materials and/or containers used to store, transport or ship blood or other potentially infectious materials. (Exception: blood, blood components or blood products released for transfusions or other clinical use.)
   B. Labels contain the "Biohazard" symbol, which is florescent orange or orange red, with letters, or symbols in a contrasting color.
   C. Labels are attached to the container with string, wire adhesive or other methods that prevent loss or unintentional removal.
   D. Red bags or red containers may be substituted for labels.
   E. Individual containers of blood or other potentially infectious materials that are placed in a pre-designated container during storage, transport, shipment or disposal are exempted from the labeling requirement.

F.  All employees are trained to recognize the method of identification of hazards and any alternative labeling or color-coding.

XI.  Hepatitis B Vaccine Program:In an effort to provide maximum protection from Hepatitis B infection, #ORG# offers a vaccination program to all employees who have potential occupational exposure to blood or OPIM. This program is provided at no cost to employees and under the supervision of a licensed physician or licensed health care professional. Refer staff to Occupational Health.

XII.  Employee Training:
   A.  All employees are trained on the Exposure Control Plan for Bloodborne Pathogens as follows:
      1.  At the time of initial assignment and annually thereafter.
      2.  Additional training is provided with modification or institution of new tasks or procedures, which may affect occupational exposure.
   B.  Appropriate training records are kept according to #ORG# Policy.

XIII.  Response to Exposure Incidents:
   A.  Employee exposure incidents are reported immediately using an Employee Injury Report form.
   B.  Employees are immediately evaluated in Occupational Health or the Emergency Room and appropriate treatment initiated, according to #ORG# Policy.

**PROCEDURES:**


**REFERENCES:**


**FORMS:**


**EQUIPMENT:**


**APPROVALS:**

| NAME | TITLE | DATE |
|------|-------|------|
| #APPROVER# | #APPRTITLE# | #APPRDATE# |

**Policy and Procedures**

Policy#: #POLNUM#
Location: #LOC#
Originating Department: #ORIGDEPT#
Effective Date: #EFFDATE#
Expiration Date: #EXPDATE#

## TITLE: EXPOSURE TO BLOOD OR OTHER POTENTIALLY INFECTIOUS MATERIALS (OPIM)

### POLICY STATEMENT:

It is the policy of #ORG# to provide and/or recommend appropriate preventive measures in accordance with the Center for Disease Control (CDC) guidelines to anyone who experiences exposure to an infectious disease while in the environs of #ORG#.

### INTENT AND SCOPE:

The policy is intended to provide guidelines for the provision of prophylactic treatment to health care workers (HCW) or others who have been exposed to bloodborne pathogens and/or other potentially infectious body materials while in the environs of #ORG#. This policy applies to students, volunteers and visitors for initial treatment and follow-up through employee health, and employees (contract and non-contract) of #ORG#.

### DEFINITIONS:

I. Health Care Worker (HCW)—Refers to all paid and unpaid persons working in the health care setting. HCWs may include, but are not limited to, physicians, nurses, aides, technicians, laboratory and morgue workers, emergency medical service (EMS) personnel, dental workers, students, volunteers, dietary workers, environmental service employees, plant operations workers, clerical staff.

II. Other Potentially Infectious Materials: (OPIM)—includes semen, vaginal secretions, cerebrospinal fluid, synovial fluid, pleural fluid, pericardial fluid, amniotic fluid, breast milk, and tissue from an HIV antibody positive source.

III. Post Exposure Prophylaxis (PEP)—Chemoprophylaxis as recommended by the Center for Disease Control (CDC) for those who have been exposed to HIV or Hepatitis B.

IV. Visitors—All others on #ORG# property for whatever reason. Includes those who are employees, but who are voluntarily attending #ORG# provided in-service and/or training not related to work.

V. Initial Treatment for non-employees—Initial treatment is defined as the first day's dose (first 24 hours) of required PEP.

### GENERAL INFORMATION:

I. It is the responsibility of the exposed HCW to immediately report an occupational health exposure to the supervisor or designee (person in "charge").

### PROCEDURES:

I. Exposure to Blood or OPIM:
   A.   Wash site thoroughly. Use Standard First Aid for eye exposure (flush with water for fifteen minutes)

B. Obtain source name and medical record number.

C. Immediately report the exposure to the supervisor or the supervisor designee (person in "charge").

   1. Employees complete an Employee Injury Report Form.

   2. All contract workers complete the forms provided for reporting and follow-up.

D. The Supervisor or designee immediately initiates the following process:

   1. #ORG# employee referred to:

      a. #ORG# Urgent Care Clinic—

      or

      b. Emergency Room—after hours.

   2. Non #ORG# employees or other individuals referred to:

      a. their private physician

      or

      b. If they choose to receive care through #ORG#, refer the individual to the appropriate #ORG# Clinic or Emergency Services to initiate treatment.

   3. An Employee Injury Report and Risk Management event report will be completed and delivered to employee health within twenty-four (24) hours or prior to the end of the next business day. Risk Management will receive the event report as soon as possible.

E. The employee health nurse or the Emergency Services practitioner determines the level of exposure per MMWR Guidelines.

II. #ORG# Employee Exposure to Blood or OPIM:Following the report of an occupational exposure incident, the employee is immediately provided with a confidential medical evaluation by employee health or the Emergency Room. Follow-up of the exposure is made by employee health.

A. Employee Health or their designates have access to the following during the evaluation of the exposure incident:

   1. Employee's duties as related to the exposure incident.

   2. Documentation of the route(s) of exposure and circumstances under which the exposure occurred.

   3. Results of the source individual's blood testing, if available.

   4. Medical records relevant to treatment of the employee, specifically, vaccination status.

B. Employee exposures are documented and reported on the Employee Injury Report form.

C. The source individual's blood is tested as soon as feasible and after consent is obtained in order to determine HBV, HCV and HIV status of the source. Test includes OHS-HIV (rapid screening), HBsAg and HCAb.

   1. When consent is denied for HIV testing, in accordance with #STATE# Statute § XX.XX, the source individual can be tested without consent by obtaining a court order, if an occupational exposure has occurred.

   2. When the source individual is known to be infected with documented HBV, HCV or HIV, testing is not repeated.

   3. Results of the source individual's testing will be made available to employee health and the exposed employee, in accordance with #STATE# Statute § XX.XX concerning the disclosure of the identity and infectious status of the source individual.

D. The employee is offered PEP when medically indicated. PEP is preferably initiated within one to two (1–2) hours post exposure.

   1. The employee experiencing an exposure where PEP is categorized as "Not Warranted" is referred for follow up to employee health. The exposed person is directed to report to employee health on the first business day post exposure.

   2. PEP is provided to those employees that are determined to have an exposure categorized as "Consider" or "Recommend" per MMWR Guidelines, subject to the employee's informed consent.

   3. If the employee consents to be placed on PEP medication, and the source is not being treated with antiretroviral medication, the treating area will obtain recommended PEP regiment: Combivir and/or Indinavir/Viracept.

   4. If source is known HIV + and has taken antiretroviral medications, the treating physician will either call the infectious disease doctor on call or the National PEP Hotline, (1-888-448-4911), for consultation.

5. In the event PEP is contraindicated or refused, the employee is directed to report to employee health for appropriate monitoring.

6. Before the employee is placed on any PEP medications, the employee must have lab drawn for: CBC w/Diff., Chem 14, Chem Liver, Pregnancy (if female, must be negative before starting PEP meds), Amylase and Lipase testing.

7. Initial dose of PEP (when indicated) is provided within two (2) hours post exposure. A prescription is written for one to three (1–3) day courses (to provide doses until next business day). Amount of medication to be provided is to continuously cover exposed employee until prescription can be obtained next day from a pharmacy.

8. The employee obtains refill prescriptions as needed.

E. The exposed employee's blood is tested by employee health as soon as feasible after consent is obtained for Hep. B, Hep. C and HIV. HIV testing can be done anonymously through employee health.

1. If the employee consents to baseline blood collection, but does not give consent for HIV serological testing, the sample is appropriately labeled and preserved for at least ninety (90) days. If, within ninety (90) days of the exposure incident, the employee elects to have baseline testing, then the testing is done as soon as feasible.

2. Pre- and Post-Counseling is provided to all individuals being tested for HIV Per #ORG# Policy

3. All positive results are reported to Public Health by name, in accordance with State Law.

4. The employee is placed on a protocol for follow up testing if needed: at two (2) weeks (if on PEP), six (6) weeks, twelve (12) weeks and six (6) months.

F. The employee receives counseling and evaluation of reported illnesses related to the exposure incident.

G. The exposed employee is provided with a copy of the evaluation within fifteen (15) days. The written opinion includes the following information:

1. Results of the medical examination.

2. Medical conditions resulting from exposure to blood or OPIM which require further evaluation or treatment.

3. All other findings or diagnosis remain confidential and are not included in the written report.

H. Employee health maintains medical records of all exposure incidents, post-exposure follow-up, and hepatitis B vaccination status. These records are kept confidential and retained for the duration of the employment plus 30 years. Medical records of employees who have terminated employment are labeled as an employee exposure and maintained by the Medical Record Department. The record includes:

1. Employees' name and social security number.

2. A copy of the employee's hepatitis B vaccination status.

3. A copy of all results of examinations, medical testing, health care professional written opinion and follow-up procedures related to an employee exposure.

4. A copy of all information related to exposure incident.

5. Employee health records are:

    a. Kept confidential.

    b. Kept separate from employment records in Human Resources.

    c. Not disclosed without the employee's expressed written consent to any person within or outside the workplace except as required by law.

III. Employee Health Follow Up Responsibilities:

A. Facilitating that the appropriate follow up of an employee with a substantiated occupational exposure to an infectious disease is completed. All communication with employee regarding exposure will be documented in the employee's medical record, including contact for follow up testing and compliance related to follow up testing. This includes reporting only Western Blot confirmed HIV positive lab results on source or employee to Public Health. Prior to lab confirmation, it is reported as "not complete, pending lab".

B. Investigation of the incident by collecting available pertinent information.

C. Consulting with Infection Control staff, as needed.

D. Presenting the information to the Medical Director for evaluation.

E. Following up on appropriate PEP and control measures.

F. Maintaining appropriate records of all employee exposures.

G. Reporting trends to the Quality and Safety Committee and other official committees as appropriate at least quarterly.

**REFERENCES:**

**FORMS:**

**EQUIPMENT:**

**APPROVALS:**

| NAME | TITLE | DATE |
|------|-------|------|
| #APPROVER# | #APPRTITLE# | #APPRDATE# |

**Policy and Procedures**

Policy#: #POLNUM#
Location: #LOC#
Originating Department: #ORIGDEPT#
Effective Date: #EFFDATE#
Expiration Date: #EXPDATE#

## TITLE: EXPOSURE TO HUMAN IMMUNODEFICIENCY VIRUS (HIV)

### POLICY STATEMENT:

It is the policy of #ORG# to provide and/or recommend appropriate preventive measures in accordance with the Center for Disease Control (CDC) guidelines to anyone who experiences exposure to an infectious disease while in the environs of #ORG#.

### INTENT AND SCOPE:

The policy is intended to provide guidelines for the provision of prophylactic treatment to health care workers (HCW) or others who have been exposed to HIV by blood and/or body fluid exposure while in the environs of #ORG#.

This policy applies to all full-time and part-time employees of #ORG# for initial treatment and follow-up through Employee Health. It also applies to contracted employees (i.e., #STATE# Medical Group and Environmental Services), students, volunteers and visitors for initial treatment by #ORG# with follow-up being the responsibility of the exposed individual unless otherwise stated in the contract, if the individual is a contract employee.

### DEFINITIONS:

I. Health Care Worker (HCW)—Refers to all paid and unpaid persons working in the health care setting. HCWs may include, but are not limited to, physicians, nurses, aids, technicians, laboratory and morgue workers, emergency medical service (EMS) personnel, dental workers, students, volunteers, dietary workers, environmental service employees, plant operations workers, clerical staff.

II. Other Potentially Infectious Materials (OPIMs)—includes semen, vaginal secretions, cerebrospinal fluid, synovial fluid, pleural fluid, pericardial fluid, amniotic fluid, breast milk, and tissue from an HIV antibody positive source.

III. Post Exposure Prophylaxis (PEP)—Chemoprophylaxis as recommended by the Center for Disease Control (CDC) for those who have been exposed to HIV.

IV. Visitors—All others on #ORG# property for whatever reason. Includes those who are employees, but who are voluntarily attending #ORG# provided in-service and/or training not related to work.

V. Initial Treatment for non-employees—Initial treatment is defined as the first day's dose (first 24 hours) of required PEP. If the incident occurs during the weekend hours between Friday night and Sunday the exposed individual is given sufficient dosage packets to continue prescribed treatment until the Monday following the exposure.

### GENERAL INFORMATION:

I. It is the responsibility of the exposed HCW to immediately report an occupational health exposure to the supervisor or designee (person in "charge") and to Employee Health or Emergency Services.

II. If the exposed employee notifies supervisory staff prior to Emergency Services or Employee Health, it is the responsibility of the supervisory staff to notify Employee Health and/or Emergency Services.

**PROCEDURES:**

I. Exposure To HIV:
   A. Wash site thoroughly. Use Standard First Aid for eye exposure (flush with water for fifteen minutes)
   B. Immediately report the exposure to the supervisor or the supervisor designee (person in "charge").
      1. Employees, including House Staff, complete an Employee Injury Report Form.
      2. All others (with the exception of Environmental Services) complete an Occurrence Report.
      3. Environmental Service employees complete a #STATE# Worker's Compensation Form.
   C. The Supervisor or designee immediately initiates the following process:
      1. Refers the employee to:
         a. Employee Health or
         b. Emergency Services—on off shifts and weekends.
      2. Refers non #ORG# employees or individuals to their private physician, or, if they choose to receive care through #ORG#, to the appropriate Community Health Center or Emergency Services to initiate treatment.
      3. Completes the Injury Report or Occurrence Report and delivers it to Employee Health or Risk Management within twenty-four (24) hours or prior to the end of the next business day.
   D. The Employee Health Nurse or the Emergency Services practitioner determines the level of exposure per MMWR Guidelines.

II. Follow Up Through Emergency Services:
   A. The exposed individual reports to the Triage Nurse with the completed Employee Injury Report/Occurrence Report/#STATE# Worker's Compensation Form in hand.
   B. The Triage Nurse notifies the practitioner nurse of an exposure.
   C. The practitioner refers to #ORG# Protocol for specific guidelines and offers appropriate PEP following exposures to HIV per status determination.
   D. All #ORG# employees with exposures that do not require PEP are reported to Employee Health. The exposed person visits the Employee Health Nurse the first business day following the exposure.

III. #ORG# Employee Exposure to Blood or OPIM:Following the report of an occupational exposure incident the employee is immediately provided with a confidential medical evaluation by Employee Health or Emergency Services (during off shifts and weekends). Follow-up of the exposure is made by Employee Health Services.
   A. Employee Health Services has access to the following during the evaluation of the exposure incident:
      1. Employee's duties as related to the exposure incident
      2. Documentation of the route(s) of exposure and circumstances under which the exposure occurred
      3. Results of the source individual's blood testing, if available
      4. All medical records relevant to treatment of the employee including vaccination status
   B. Employee exposures are documented and reported on the Employee Injury Report form.
   C. The source individual's blood is tested as soon as feasible and after consent is obtained in order to determine HBV, HCV and HIV status of the source. (Test includes OSH-HIV (rapid screening), Hep B sAg and Hep C AB.)
      1. When consent is denied for HIV testing, in accordance with #STATE# Statute § XX.XX, the source individual can be tested without consent by obtaining a court order, if an occupational exposure has occurred.
      2. When the source individual is already known to be infected with HBV, HCV or HIV testing is not repeated.
      3. Results of the source individual's testing is made available to Employee Health Services and the exposed employee, in accordance with #STATE# Statute § XX.XX concerning the disclosure of the identity and infectious status of the source individual.
   D. The employee is offered PEP when medically indicated. PEP is preferably initiated within one to two (1–2) hours post exposure.
      1. PEP is provided to those employees that are determined to be exposure categorized as "Offer" per MMWR Guidelines, subject to the employee's informed consent.

2. The employee experiencing exposure categorized as "Not Offered" is referred for follow up to the Employee Health Services office. The exposed person is directed to report to Employee Health on the first business day post exposure.

3. The initial dose of PEP (when indicated) is provided at the time of initial contact and a prescription for adequate dosage is given to last until the employee is seen by Employee Health.

4. The employee obtains refill prescriptions as needed from Occupational Health.

5. PEP is discontinued as soon as a negative HIV report is returned on the source.

6. In the event PEP is contraindicated or refused, the employee is directed to report to Employee Health for appropriate monitoring.

E. The exposed employee's blood is tested as soon as feasible after consent is obtained.

1. If the employee consents to baseline blood collection, but does not give consent for HIV serological testing, the sample is appropriately labeled and preserved for at least 90 days. If, within 90 days of the exposure incident, the employee elects to have baseline testing, then the testing is done as soon as feasible.

2. Pre- and Post-Counseling is provided to all individuals being tested for HIV per #ORG# Policy.

3. All positive results are reported to Public Health by name, in accordance with State Law.

4. The employee is placed on a protocol for HIV testing: at 6 weeks, 12 weeks and 6 months.

5. If the source individual tests negative for HIV then PEP is discontinued.

F. The employee receives counseling and evaluation of reported illnesses related to the exposure incident.

G. The exposed employee is provided with a copy of the evaluation within fifteen (15) days. The written opinion includes the following information:

1. Results of the medical examination.

2. Medical conditions resulting from exposure to blood or other potentially infectious materials which require further evaluation or treatment.

3. All other findings or diagnosis remain confidential and are not included in the written report.

H. Employee Health Services maintains medical records of all exposure incidents, post-exposure follow-up, and hepatitis B vaccination status. These records are kept confidential and retained for the duration of the employment plus 30 years. Medical records of employees who have terminated employment are labeled as an occupational exposure and maintained by the Medical Record Department. The record includes:

1. Employees' name and social security number.

2. A copy of the employee's hepatitis B vaccination status.

3. A copy of all results of examinations, medical testing, health care professional written opinion and follow-up procedures related to an employee exposure.

4. A copy of all information related to exposure incident.

5. Employee records are:

    a. Kept confidential.

    b. Not disclosed without the employee's expressed written consent to any person within or outside the workplace except as required by law.

IV. Employee Health Follow Up Responsibilities:

A. Ensuring that the appropriate follow up of an employee with a substantiated occupational exposure to an infectious disease is completed. This includes reporting only Western Blot confirmed HIV positive lab results on source or employee to Public Health. Prior to lab confirmation, it is reported as not complete, pending lab.

B. Investigation of the incident by collecting available pertinent information.

C. Consulting with Infection Control staff, as needed.

D. Presenting the information to the Medical Director for evaluation.

E. Following up on appropriate PEP and control measures.

F. Maintaining appropriate records of all employee exposures.

Reporting information to the Quality and Safety Committee and to Infection Control staff at least quarterly.

**REFERENCES:**

**FORMS:**

**EQUIPMENT:**

**APPROVALS:**

| NAME | TITLE | DATE |
|------|-------|------|
| #APPROVER# | #APPRTITLE# | #APPRDATE# |

**Policy and Procedures**

Policy#: #POLNUM#
Location: #LOC#
Originating Department: #ORIGDEPT#
Effective Date: #EFFDATE#
Expiration Date: #EXPDATE#

## TITLE: FALL PREVENTION PROGRAM

### POLICY STATEMENT:

It is the policy of #ORG# to successfully recognize patients who are at risk for falls, to prevent patient falls and protect patients from injury, and to improve patient safety.

### INTENT AND SCOPE:

It is the intent of this policy to provide guidelines for patient management to reduce patient falls.

This policy affects all Medical staff, Registered Nurses (RN), Graduate Nurses (GN) and specially qualified Licensed Vocational/Practical Nurses (LVN/LPN) and other unlicensed assistive personnel (contract or non-contract) of the #ORG# who meet qualifications.

### DEFINITIONS:

I. Fall: A sudden, uncontrolled, unintentional, descent to the floor (or extension of the floor, e.g., trash can or other equipment) with or without injury to the patient.

II. Assisted fall: When a staff member was with the patient and attempted to minimize the impact of the fall by easing the patient's descent to the floor or in some manner attempting to break the patient's fall.

### GENERAL INFORMATION:

I. To determine fall risk, patients are assessed by licensed nurses using the Morse Fall Scale (MFS)
   A. At the time of admission for care and treatment
   B. Review risk level as required based on assessment and reassessment process
   C. Review risk level with any change in patient condition or level of care

II. Based upon this assessment, the patient's fall risk are assigned as
   A. Low (Level I) Score of 20 or less
   B. Moderate (Level II) Score of 25–45
   C. High (Level III) Score of 50 or more—for outpatient care settings, determine if this is appropriate for the patients with high risk fall scores. May require an inpatient setting for surgery or other procedures.

### EQUIPMENT/SUPPLIES:

   A. Wrist bands that designate fall risk;
   B. Fall alert signs;
   C. Non-skid slippers;
   D. Bed alarms; and
   E. Other equipment as indicated.

III. Protocol:
  A. Use the Falls Risk Assessment Tool (Morse Fall Scale) as directed in the procedure section,
  B.  Document the Fall Risk Assessment Score on the designated area of the patient's chart.
  C. Implement interventions that correspond with the patient risk level and individual patient needs (see Morse Fall Scale).

## PROCEDURES:

I. Level I (MFS 20 or less) Appropriate Interventions:
  A. Reduce environmental hazards (i.e. wipe up spills, clear path to bathroom, check that patient has non-skid, adequately fitting, low-heeled footwear, and if not, order non-skid sock; check that clothing fits adequately and is not likely to trip patient while ambulating).
  B. Orient patient to room and bathroom location. Re-orient as necessary. Assist patient ambulating as appropriate.
  C. Maintain bed in low position with brakes locked.
  D. Use brakes on wheelchair/gerichair/stretcher to maintain stability whenever not moving.
  E. Validate that patient knows how and is able to use call light, call light is in reach and call light works.
  F. Validate that personal items that are requested are in reach (i.e. glasses, urinal, water, dentures, tissues, cosmetic kit).
  G. Use side rails as appropriate.
  H. Conduct safety checks for the bed in low position or when patient is up in a chair; call light within reach and a clear path to the bathroom.

II. Level II (MFS 25–45) Appropriate Interventions include but are not limited to:
  A. Include all Level I interventions, plus;
  B. Use a color identification system to identify the patient at high risk for falls according to the following;
    1. Place a wrist band on the patient.
    2. Place a fall alert sign on the patient's door.
    3. Place double treaded socks that match the wrist band for falls on patient.
  C. Inform patient/family of risk for falling and interventions being used.
  D. Instruct patient/family to call for assistance when the patient wants to get out of bed. Frequently reinforce.
  E. Evaluate the number of staff needed to get the patient OOB and to ambulate safely.
  F. Check patient at least every 2 hours to see if patient needs anything (i.e. help with toileting, hygiene assistance, etc.).
  G. Use a gait belt to assist during transfers/ambulation as appropriate.
  H. Instruct patient to get up slowly from a lying position.
  I. Review medication profile for potential side effects such as drowsiness, muscle weakness, orthostatic hypotension, confusion, or dizziness.

III. Level III (MFS 50 or greater) Appropriate Interventions include all level one and level two interventions, plus:
  A. Assign patient to a room as close as possible nurses' station.
  B. Stay in close visual or verbal contact when patient is on bedside by the commode, in bathroom, shower or at sink.
  C. Consider patient sitter.
  D. Seek cooperation from patient's family to stay with the patient at all times. Instruct patient/family to call for help when the patient wants to get out of bed.
  E. Low bed position.

IV. Patient/Family Education
  A. Educate the patient/family at the level of their understanding the following preventive measures to decrease the potential of falls:
    1. The reason for fall prevention measures, when used.
    2. Decreasing environmental risks.
    3. The need to ask for assistance when getting out of bed.

V. Documentation
    A.   For moderate and high fall risk patients, document interventions implemented.

**REFERENCES:**

**FORMS:**

**EQUIPMENT:**

**APPROVALS:**

| NAME | TITLE | DATE |
|------|-------|------|
| #APPROVER# | #APPRTITLE# | #APPRDATE# |

**Policy and Procedures**

Policy#: #POLNUM#
Location: #LOC#
Originating Department: #ORIGDEPT#
Effective Date: #EFFDATE#
Expiration Date: #EXPDATE#

## TITLE: FIRE SAFETY MANAGEMENT PLAN

### POLICY STATEMENT:

It is the policy of #ORG# to utilize a Fire Safety Management Plan in order to provide a fire-safe environment of care for the protection of our patients, employees, physicians, volunteers, visitors and property.

### INTENT AND SCOPE:

The purpose of this policy is to define standardized mechanisms to conduct business and to meet legal and regulatory requirements for providing Fire Safety Management procedures for staff. This policy applies to volunteers, Medical Staff, employees (contract and non-contract) and property of #ORG#. The program maintains all applicable required structural features of fire protection of the Life Safety Code (LSC) (NFPA 101####edition). Performance is measured, in part, through the establishment of standards that will measure staff knowledge, critique fire drills, quality control monitors, incident reporting, and equipment testing and inspection. Objectives are:

1. Ensure proper operation of fire detection, alarm and suppression systems through a quality control program that oversees inspection, testing and maintenance.
2. Provide portable fire extinguishers that are appropriate for the facility.
3. Utilize curtains, furniture, waste baskets, bedding and other equipment that meets appropriate fire safety criteria.
4. Investigate and act on deficiencies, failures and errors in the system.
5. Establish a training and education program for employees, contract workers, physicians and practitioners that are new, with ongoing annual training thereafter.
6. Provide emergency response processes for the facility that address unit specific needs, fire evacuation routes, and assigned roles and responsibilities that address proximity to the fire, away from the fire and evacuation.
7. Continually analyze data from drills, training, knowledge and awareness, equipment response and skill during fire emergencies, facility code compliance and other fire issues, correcting deficiencies rapidly and completely.

### DEFINITIONS:

NFPA: National Fire Protection Agency.
NFPA 101: The Life Safety Code is the registered trademark of an American consensus standard which, like many NFPA documents, is methodically updated every three years. The standard is not a legal code but is adopted by various organizations for legal use. This code coincides with building, fire and sanitation codes for example.
RACE: Rescue, Alarm, Contain, Extinguish
Statement of Conditions (SOC): The Joint Commission (TJC) requires Health Care organizations to document through the use of Statement of Conditions (SOC) that a fire safe environment of care is maintained and that the facility complies with the intent of NFPA 101 Life Safety Code.

**GENERAL INFORMATION:**

I. Unlike other industries, where escape and evacuation may be the first response to a fire, #ORG# buildings are constructed to protect and aid staff with incipient fire response measures because patients in general are not able to protect themselves and escape from danger.

II. The goal of the Fire Safety Management plan is to establish and maintain a fire protection program to promote a safe, controlled, and comfortable environment by:
   A. Providing operational reliability of fire protection systems.
   B. Reducing the potential for fire through education and prevention.
   C. Minimizing potential risks of fire protection system failures by conducting scheduled and prevention maintenance.

III. The Quality and Safety Committee is responsible for developing, implementing, and monitoring fire safety performance regulations, standards and processes for the #ORG#.

**PROCEDURES:**

I. The #ORG# policy for Fire Safety Management Plan is developed, maintained and assessed, describing the processes implemented to manage fire safety risks.

II. Developing and enforcing #ORG# smoking policy, performing maintenance on electrical equipment, conducting new hire and ongoing fire safety training and education, maintaining smoke and heat detection systems, and maintaining fire alarm and suppression systems is incorporated into the plan and process.

III. The inspection, testing, and maintenance of all fire protection equipment are the responsibility of the designated administrator via plant operations contracted services. Scheduled downtimes and the anticipated length of downtimes of alarm, detection, or extinguishing systems are announced in advance to all affected departments.

IV. The #ORG# Disaster Plan has been developed to provide staff with procedures required in responding to a fire.
   A. Hard copy versions of the Disaster Plan are available in red-colored binders titled #ORG# Disaster Plan and electronic versions of the same plan are available online.

V. Departments acquiring bedding, window draperies, furnishings, decorations, and other equipment are to exercise an assessment for flammability labeling or certifications declaring the products feature resistance for fire development or spreading to minimize fire vulnerability. Bedding, draperies, and furnishings are generally affixed with a flammability label or tag certifying products have been tested for fire resistance. In addition, vendors of furniture can provide information regarding fire safety certifications.
   A. Combustible decorations are prohibited unless one (1) of the following criteria is met:
      1. The decoration is flame-resistant; or,
      2. Decorations such as photographs and paintings are in such limited quantities that a hazard of fire development or spread is not present.

VI. The Statement of Conditions (SOC) documents, or other accrediting/regulatory body equivalent, are used to demonstrate compliance with the intent of the National Fire Protection Association Life Safety Code standards.
   A. Administration and Plant Operations is responsible for completing and maintaining record of an annual review of these documents.

VII. The SOC documents, or other accrediting/regulatory body equivalent, require ongoing management to continually identify and assess Life Safety Code deficiencies.
   A. Administration and Plant Operations utilizes the appropriate accrediting body tool to report and correct Life Safety Code deficiencies.

VIII. The corrective plan outlines a plan of action to correct Life Safety Code deficiencies, and establishes dates for corrections to be completed. Administration and Plant Operations is responsible for continual improvement until completion of the corrective plan is approved by regulatory and accrediting bodies or internally.

IX. Administration and Plant Operations is responsible for the completion of SOC or equivalent document. Outsourcing of activity to person(s) with considerable education, training, or experience of the Life Safety Code is permitted.

X. Fire drills are conducted quarterly for the ambulatory center. Fire drill quarter groupings are: January through March, April through June, July through September, and October through December. At least 50% of the required drills are unannounced.

    A. Fire drill exercises are not pre-announced to staff and are designed to develop knowledge needed to support high reliability and safe rapid responses to a fire emergency.

    B. Staff in all areas of every building where patients are housed or treated participates in drills.

    C. Staff roles and responsibilities in a fire's point of origin and away from a fire's point of origin are defined in #ORG# Disaster Plan's Fire Emergency Action Plan.

        1. Hard copy versions of the Disaster Plan are available in red colored binders labeled Disaster Plan. Electronic versions of the Disaster Plan are available online.

        2. All fire drills are critiqued to identify deficiencies and opportunities for improvement.

        3. Fire drills are critiqued by grading on a 1 to 3 scale for staff actions on important fire event response measures.
Score 3: Staff response reflected a thorough understanding of the RACE response subject.
Score 2: Staff response reflected an adequate understanding of the RACE response subject.
Score 1: Staff response reflected a limited or no understanding of RACE response subject.

    D. An annual assessment of #ORG#'s fire response training as well as evaluation of other Environment of Care functions is provided to the Quality and Safety Committee and to #ORG# leadership.

    E. During fire drills, staff knowledge is evaluated.

    F. Evaluation of staff knowledge in fire drills is documented on form titled, "#ORG# Code Red Response Evaluation Form". The form enables grading of staff response for the following:

        1. When and how to sound fire alarms, where such alarms are available.

        2. When and how to transmit for offsite fire responders.

        3. Containment of smoke and fire.

        4. Transfer of patients to areas of refuge.

        5. Fire extinguishment.

        6. Specific fire response duties.

        7. Preparation for building evacuation.

    G. The Plant Operations personnel are responsible for testing initiating devices and fire detection and alarm equipment in accordance with NFPA 72. Testing intervals for initiating devices is as follow:

        1. All supervisory signal devices (except valve tamper switches) are tested at least quarterly.

        2. All valve tamper switches and water flow devices are tested at least semiannually.

        3. All duct detectors, electromechanical releasing devices, heat detectors, manual fire alarm boxes, and smoke detectors are tested at least annually.

    H. Occupant alarm notification device testing is performed by certified contractor and documentation of testing is maintained in the Plant Operations Department.

    I. Water-based automatic fire-extinguishing system testing is performed by certified contractor and documentation of testing is maintained in the Plant Operations Department.

    J. Water-based automatic fire-extinguishing system testing scope is:

        1. All fire pumps are tested at least weekly under no-flow condition.

        2. Main drain tests are conducted at least annually at all system risers.

        3. All fire department connections are inspected quarterly.

        4. All fire pumps are tested at least annually under flow.

    K. Kitchen automatic fire extinguishing system testing is performed by certified contractor and documentation of testing is maintained in administration.

    L. Semi-annual inspection of carbon dioxide and other gaseous automatic fire extinguishing systems is performed by certified contractor under the administration of Plant Operations and documentation of inspection is maintained in administration.

M. Annual maintenance of portable fire extinguishers is performed by certified contractor in accordance with NFPA 10 and records of inspection and maintenance activities for fire extinguishers are maintained in administration.

N. If hoses are attached to standpipes, the inspection records would be kept in administration.

O. Documentation of fire and smoke damper maintenance and testing activity is maintained in administration.

P. Documented results from testing of automatic smoke-detection shutdown devices for air-handling equipment are maintained in administration.

Q. Horizontal and vertical sliding and rolling doors are kept closed at all times or have an automatic release system that closes the door when the fire alarm system is activated. Doors are not propped or blocked open or tampered with in such a way as to prevent automatic closure.

R. Refer to #ORG# policy, Interim Life Safety Measures for additional guidance.

**REFERENCES:**

**FORMS:**

**EQUIPMENT:**

**APPROVALS:**

| NAME | TITLE | DATE |
| --- | --- | --- |
| #APPROVER# | #APPRTITLE# | #APPRDATE# |

**Policy and Procedures**

Policy#: #POLNUM#
Location: #LOC#
Originating Department: #ORIGDEPT#
Effective Date: #EFFDATE#
Expiration Date: #EXPDATE#

## TITLE: HAND HYGIENE

### POLICY STATEMENT:

It is the policy of #ORG# that health care workers practice appropriate hand hygiene in the healthcare setting.

### INTENT AND SCOPE:

This policy is intended to provide guidelines to reduce the transmission of pathogenic microorganisms to patients and personnel in the healthcare setting. The policy applies to students, volunteers, Medical Staff and employees (contract and non-contract) of #ORG#.

### DEFINITIONS:

I. Hand Hygiene—includes a waterless antiseptic hand rub, exogenous water not required, plain soap or anti-microbial soap and water, and surgical hand antisepsis.

II. Decontaminate hands—Use of an alcohol-based waterless antiseptic agent.

III. Artificial Fingernails—Fingernails that are not natural of any type, sculptured or silk wrap, gel, bonded, separate, press on, complete or "tips".

IV. High Risk Patients/Areas-Patients that is vulnerable to infections.

V. Direct Patient Caregivers—Healthcare Workers (HCWs) who provide care to the patient at the bedside.

### GENERAL INFORMATION:

### PROCEDURES:

I. If hands are not visibly soiled an alcohol based waterless antiseptic agent is used in all clinical situations (except patients with C-Diff) as described:
  A. After contact with a patient skin (example: taking B/P, pulse, turning, lifting)
  B. After contact with body fluids or excretions, mucous membranes, non-intact skin, or wound dressings, as long as hands are not visibly soiled.
  C. If moving from a contaminated body site to a clean body site during patient care.
  D. After contact with inanimate objects (including medical equipment) in the vicinity of the patient.
  E. Before caring for patients with severe neutropenia or other forms of severe immune suppression.
  F. Before invasive procedures and after removing gloves.
  G. Before insertion of devices that do not require a surgical procedure.

II. When decontaminating hands with a waterless antiseptic:
    A. Apply to the palm of hand
    B. Rub hands together covering all surfaces of hands and finger until hands are dry, usually ten to fifteen (10–15) seconds.

**NOTE:** Follow manufacturer's recommendations on the volume of product to use.

III. Artificial Fingernails: Direct patient caregivers are prohibited from wearing artificial fingernails. Recommended length of nails for ALL direct patient caregivers is no longer than one fourth ($^1/_4$) of an inch from the tip of the finger.

IV. Hands are washed with a non-antimicrobial soap and water or an antimicrobial soap and water when hands are visibly dirty or if the patient has Clostridium Difficile.

V. Hand washing:
    A. Remove jewelry,
    B. Wet hands under running water,
    C. Keeping hands lower than elbow, apply soap,
    D. Use friction to clean between fingers, palms back of hands, wrist, and forearms, clean, under nails.
    E. Wash for ten to fifteen (10–15) seconds,
    F. Rinse under running water,
    G. Use paper towels to thoroughly dry hands,
    H. Use paper a paper towel to turn off the faucet and discard.

VI. Skin Care:
    A. Lotions;
        1. Use of community or private hand lotions is discouraged.
        2. Infection Control approved dispenser lotion is made available in the healthcare setting for employee use.

VII. Information:
    A. General hand hygiene is practiced;
        1. Before starting to work,
        2. Before preparing food, water, or medication,
        3. Before and after eating,
        4. After using the toilet,
        5. Before and after removal of gloves,
        6. Wear gloves when it can be reasonably anticipated that contact with blood or other potentially infectious materials, mucous membranes, and non-intact skin occur.

**REFERENCES:**

**FORMS:**

**EQUIPMENT:**

**APPROVALS:**

| NAME | TITLE | DATE |
|---|---|---|
| #APPROVER# | #APPRTITLE# | #APPRDATE# |

**Policy and Procedures**

Policy#:  #POLNUM#
Location:  #LOC#
Originating Department:  #ORIGDEPT#
Effective Date:  #EFFDATE#
Expiration Date:  #EXPDATE#

## TITLE: HAZARDOUS MATERIALS AND WASTE MANAGEMENT PLAN

### POLICY STATEMENT:

It is the policy of #ORG# to provide a plan to comply with local, state, and federal laws in the collection, handling and disposing of all hazardous materials and wastes. The plan includes processes designed to minimize the risk of harm. The processes include education, procedures for safe use, storage and disposal, and management of spills or exposures.

### INTENT AND SCOPE:

This policy is intended to meet legal or regulatory requirements. The Hazardous Materials and Waste Management Plan's scope is to provide a program to safely control hazardous materials and waste in the environment of #ORG#. This policy applies to #ORG# employees (contract and non-contract), Medical staff and volunteers.

### DEFINITIONS:

#### GENERAL INFORMATION:

   I. Hazards associated with materials and wastes are defined in Material Safety Data Sheets (MSDS) or similar documents provided by suppliers and manufacturers.

  II. Segregation of hazardous waste at the point of generation is an effective means of controlling the potential for exposures or spills during collection, transport, storage, and disposal.

 III. Departments are responsible for orienting new personnel to the department and, as appropriate, to job and task specific uses of hazardous material or waste. All personnel are responsible for learning and following job and task specific procedures for safe handling and use of hazardous materials and waste.

#### PROCEDURES:

   I. Managing risks related to Hazardous Materials and Waste.
      A. This policy provides a plan to comply with local, state, and federal laws in the collection, handling and disposing of all hazardous materials and wastes

  II. Current inventory of hazardous materials
      A. Departments maintain an inventory of the hazardous materials and wastes they generate and are responsible for the safe selection, storage, handling, use, and disposal.
      B. Regulated Medical Waste (Red-Bag), including sharps, are picked up by environmental service department staff in patient care areas and transported to temporary storage area in sole-function carts. The waste is packaged for disposal and held for pickup by a licensed waste contractor. The contractor assists in completing the manifest and removes the waste, returning the disposal copy of the manifest after disposal of waste at treatment facility.

III. Hazardous materials and waste spills or exposures procedures.
  A. #ORG#'s spill procedure evaluates spills to determine if outside assistance is necessary. A minor (incidental) spill is defined as one that can be cleaned up by the staff with use of personal protective equipment provided as part of their employment. A major spill is defined as one that exceeds the capability of the staff to neutralize and clean up, and requires a response from emergency forces (fire department). In these cases, the area is evacuated, ventilation controlled, and the Fire Department's HAZMAT Team notified. The Fire Department takes control of the site and cleans up or can arrange for it to be cleaned up. Staff is cautioned to not to handle chemical spills that exceed their training or the personal protection available.

IV. Implementing procedures in response to hazardous material and waste spills or exposures.
  A. The emergency spills process is initiated per policy.

V. Minimizing risk for waste materials.
  A. Infectious and regulated medical wastes including sharps are common waste products throughout #ORG#. A licensed independent contractor collects waste for transport and disposal at an approved off-site facility.
  B. The Hazardous Waste Coordinator (Safety Officer) assesses the appropriateness of space for handling and storage of hazardous materials and waste as part of the environmental tour program. The goal in evaluating these issues is to determine if current conditions and practices support safe handling and storage of hazardous materials and waste, and separation of the hazardous waste from clean and sterile goods. Departments are responsible for initiating corrective actions on findings related to the appropriate use of handling and storage spaces in their areas of responsibility.
  C. Department leaders are responsible for taking measures to minimize risks associated with hazardous materials and waste by:
    1. Evaluating Material Safety Data Sheets for hazards before purchase of departmental supplies to ensure they are appropriate and the least hazardous practical.
    2. Providing equipment for the safe storage and handling of hazardous material and waste.
    3. Maintaining data sheets to provide hazard warnings associated with hazardous material and waste.
    4. Managing the proper disposal of hazardous waste generated.

VI. Minimizing risks for chemotherapeutic waste.
  A. Department Management and Environmental Services share responsibility for the disposal of chemotherapeutic medical waste through licensed contractors who transport this waste.

VII. Minimizing risk for hazardous energy sources.
  A. Reference radiation safety policy establishes standard procedures for radiation safety to comply with #STATE# Department of Health and Office of Radiation Control. Materials and waste are handled in accordance with #ORG#'s Nuclear Regulatory Commission License and safety is managed by the Radiation Safety Officer.

VIII. Minimize risks associated with disposing of hazardous medications.
  A. Reference #ORG# policy, Documentation of Controlled Substances. Policy establishes processes for documentation of controlled substance drug wastage.

IX. Minimize risks associated with selection, handling, storage, transport, use, and disposing hazardous gases and vapors.
  A. Monitoring of gases and vapors, gases and vapors is performed annually. In all locations where anesthesia is administered, engineering controls such as a scavenging system to remove waste anesthetic gases and adequate room ventilation are utilized to minimize risks.

X. Monitoring levels of hazardous gases and vapors to determine that they are in safe range.
  A. If results reflect exposure levels above the permissible exposure level, corrective action and additional testing will be done to demonstrate a safe working environment. The results are reported to the affected departments and the Quality and Safety Committee at least annually.

XI. Permits, licenses, manifests, and material safety data sheets required by law and regulation are available.
   A. #ORG# maintains permits and licenses for handling and disposing of hazardous wastes, including chemical wastes, radioactive materials, and potentially infectious medical wastes from the appropriate federal, state, and municipal agencies.

XII. Labeling hazardous materials and waste.
   A. All hazardous materials and wastes are labeled specific signs or with text labels in accordance with Federal standards.
      1. Infectious Waste: These wastes are items that have the potential of infecting another individual. This type of waste is placed in red bags labeled, "Hazardous Waste". Items placed in red bags include contaminated tubing, soiled dressings, and other disposables contaminated with blood and body fluids. The red bags are not used for any other purpose and any material in a red bag is treated as infectious waste.
      2. Infectious Bio-hazardous Waste: These wastes are objects, i.e., needles, glass, blades, etc., capable of puncturing the skin, such as needles and blades. This type of waste is placed in red impervious and sealed plastic containers made of puncture resistant material labeled "Hazardous Waste".
      3. Chemotherapeutic Waste: Chemotherapeutic wastes are carcinogenic wastes. These wastes are disposed in yellow-colored bags with additional warning labels.
      4. Chemical Materials and Waste: Chemical materials are labeled throughout their use and handling. Hazard warning labels are generally on the container prior to receipt or labeling of empty blank container is done when filled for use in smaller quantity, in accordance with NFPA 704, Standard System for the Identification of the Hazards of Materials for Emergency Response. The standard identifies hazards and their severity using a numerical method to describe the relative hazards of a material.
         a. Chemical wastes are labeled on the containers. These labels are required by the contractors of chemical disposal services to maintain the identity of the materials. If the identity is lost of the chemical waste is lost, the materials are tested and analyzed to identify them for proper handling and disposal.
      5. Radioactive Materials and Waste: Radioactive materials are labeled with the trefoil sign (three leaf plant) and the sign is black against a yellow background. Radioactive materials are handled and stored in accordance with the NRC regulations and license provisions.

**REFERENCES:**

**FORMS:**

**EQUIPMENT:**

**APPROVALS:**

| NAME | TITLE | DATE |
|------|-------|------|
| #APPROVER# | #APPRTITLE# | #APPRDATE# |

**Policy and Procedures**

Policy#: #POLNUM#
Location: #LOC#
Originating Department: #ORIGDEPT#
Effective Date: #EFFDATE#
Expiration Date: #EXPDATE#

## TITLE: HISTORY AND PHYSICAL EXAMINATION

### POLICY STATEMENT:

It is the policy of #ORG# that all Medical Staff members practicing at #ORG# to establish standards for the history and physical (H&P) examination consistent with legal regulations and accreditation standards.

### INTENT AND SCOPE:

It is the intent of this policy to include documentation standards for the initial diagnostic evaluation of patients. It applies to all Medical staff and employees (contract and non-contract) of #ORG#. This policy contributes information and standards that lend to timely and accurate documentation for our patients, regardless of the venue of care, whether outpatient or inpatient.

### DEFINITIONS:

I. Authentication—This occurs when the author/owner of the history and physical signs, dates and times the document. This can be done electronically or by hand in the medical record. More than one qualified practitioner can participate in performing, documenting, and authenticating an H&P for a single patient. When performance, documentation, and authentication are split among qualified practitioners, the practitioner who authenticates the H&P will be held accountable for its contents.

II. History and physical exam (H&P)—The initial assessment of the patient, which documents the current and relevant prior medical history, physical examination, diagnosis or differential diagnosis, and treatment plan. The assessment may constitute a full H&P or a short form H&P depending on the admission status/procedure being performed.

III. Interval Note—A note written on admission or prior to surgery, detailing changes in the patient's history and examination, which have occurred since the H&P was completed.

IV. Procedure—An operation, treatment, or test performed in the operating room (OR) suite, outpatient clinic, ambulatory surgery center or procedural sedations areas.

V. Operative or complex invasive procedures—Procedures performed in an operating room or procedures involving anesthesia or monitored anesthesia care.

VI. Major high-risk diagnostic or therapeutic intervention—Procedures involving general anesthesia to patients who are poor anesthesia risks (for example, ASA rating greater than 3; procedures which carry a significant risk of adverse outcome, complications, or other sequela; procedures performed on patients with co-morbidity likely to adversely effect the prospects for a favorable outcome).

**GENERAL INFORMATION:**

I. Content of H&P Examination:
A. The H&P contains sufficient information to support the diagnosis or differential diagnosis, justify the treatment plan, and facilitate the care after discharge.
B. Patients requiring an H&P receive a full H&P, a focused short H&P or an interval note as set forth in this policy.
1. A full H&P is defined as one that contains the following elements:
a. Appropriate and relevant history,
b. Physical exam,
c. For children, an evaluation of the development age,
d. Other relevant elements: advance directives, informed consent,
e. Assessment and Plan,
f. Patient's condition
2. A short form H&P contains the following elements:
a. Appropriate and relevant history,
b. Physical exam,
c. Assessment and plan,
d. Patient's condition.
3. An interval note is a statement entered into the medical record that an H&P has been reviewed and that states the following:
a. There are no significant changes to the findings contained in the H&P since the time such H&P was performed, or
b. There are significant changes and such changes are subsequently documented in the medical record.
4. Documentation of the H&P examination:
a. The H&P may be dictated and transcribed, computer generated, or handwritten.
b. The H&P is legible and documented in a manner so as to be durable and permanent.

The H&P is authenticated: signed, dated and timed electronically or by hand.

**PROCEDURES:**

I. A medical history and appropriate physical examination is performed on patients being admitted for inpatient care.

II. An H&P examination is performed and documented by the following:
A. A qualified physician (doctor of allopathic or osteopathic) who is a member of the professional staff and who, by virtue of education, training, and demonstrated competence, is granted clinical privileges to perform specific diagnostic and therapeutic procedures and who is fully licensed to practice medicine in the state.
B. Oral and maxillofacial surgeons if they possess the clinical privileges to do so in order to assess the medical, surgical/anesthetic risks of the proposed operative/other procedure.
C. Dentist and podiatrists are responsible for that part of the patient's H&P that relate, respectively, to dentistry and podiatry. They may perform a complete H&P examination if they possess clinical privileges to do so. A qualified physician endorses the findings prior to any major high-risk diagnostic or therapeutic intervention.
D. Podiatrists may complete the admission and pre-operative H&P's on American Society of anesthesiologists (ASA) Class 1 and II patients. Patients in (ASA) Class III, IV and V will require a M.D. or D.O. evaluation upon admission or scheduling of outpatient procedures.
E. A Credentialed and Privileged Non-licensed Independent Practitioners or physician extender such as a Physician Assistant or Nurse Practitioner can perform the History and Physical (or Part of It) if:
1. The individual is supervised and sponsored by a licensed independent physician, podiatrist, or dentist.
2. The supervising person accepts moral and legal responsibility for the completeness and quality of the H&P.
3. The individual is appropriately credentialed and granted privileges to perform this function.

III. Timing and expiration of the H&P exam:
A. The admission H&P exam is good for the entire patient stay.

B. A complete H&P examination is performed and documented and in the record within twentyfour (24) hours after admission.

C. The H&P may be performed up to thirty (30) days prior to admission. If performed prior to admission it is updated within twentyfour (24) hours of admission. If the patient is having surgery or other procedure that places the patient at risk and or involves the use of sedation or anesthesia within the first twentyfour (24) hours of admission, there is an update to the patient's condition prior to the start of surgery.

D. The H&P examination, the results of any indicated diagnostic tests, as well as a provisional diagnosis are recorded before the operative procedure by the licensed independent practitioner responsible for the patient. The Medical Staff Executive Committee may authorize special H&P forms for certain categories of patients.

E. Except in emergency, no patient is admitted to the hospital until a provisional diagnosis has been documented in the medical record. In case of emergency, the provisional diagnosis is stated as soon after admission as possible.

F. When the H&P examination is dictated and the transcribed report is not available at the time of the operation, an admission note documenting significant H&P examination findings are recorded in the medical record(s) before the operation proceeds.

G. For ambulatory surgery centers and elected surgical or other procedures, the authenticated history and physical is on the chart prior to the surgery.

**REFERENCES:**

**FORMS:**

**EQUIPMENT:**

**APPROVALS:**

| NAME | TITLE | DATE |
|------|-------|------|
| #APPROVER# | #APPRTITLE# | #APPRDATE# |

**Policy and Procedures**

Policy#:  #POLNUM#
Location:  #LOC#
Originating Department:  #ORIGDEPT#
Effective Date:  #EFFDATE#
Expiration Date:  #EXPDATE#

## TITLE: IMPAIRED PHYSICIANS AND PRACTITIONERS

### POLICY STATEMENT:

It is the policy of #ORG# and its Medical Staff to require all physicians/practitioners to provide continuous competent care of patients and maintain adequate health status.

### INTENT AND SCOPE:

The policy is intended to provide overall guidance and direction on how to proceed when confronted with a potentially impaired practitioner.

### DEFINITIONS:

I. Impaired Physician/practitioner—a physician/practitioner who is unable to practice medicine with reasonable skill and safety to patients because of a physical or mental illness, including deterioration or loss of cognitive skills, motor skills or excessive use or abuse of drugs, including alcohol. (American Medical Association Definition)

II. Investigative Team:
   A. An Ad Hoc Committee of the Medical Staff as appointed by the Medical Director
   B. Membership:
      1. Medical Director.
      2. Two (2) physician members.
      3. Two (2) individuals appropriate upon the circumstances involved.
   C. Practitioner—Physician or mid-level provider credentialed and approved by the medical staff to provide medical care and treatment to patients of #ORG#.

### GENERAL INFORMATION:

I. Any individual working in #ORG# that has a reasonable suspicion that a practitioner with privileges may be physically or mentally impaired notifies the Medical Director of such concern.

II. Any practitioner whose health status changes in such a manner as to jeopardize his/her ability to carry out his/her privileges, or to provide safe medical care shall notify the Medical Director as soon as practical.

III. An investigation of a potentially impaired practitioner is conducted in a manner that is separate from the usual medical staff disciplinary process and is appropriate to the circumstances of the incident(s) and individuals involved.

IV. The Medical Director and the Department Chairman will meet with a practitioner upon receipt of a report from an individual or the practitioner concerning a change in his/her health status that jeopardizes the individual's ability to carry out privileges or provide safe medical care to #ORG#'s patients.

V. Confidentiality is maintained throughout the investigation. All parties involved avoid speculation, conclusions, gossip, and any discussion of the matter with anyone other than those individuals participating in the process described in this policy.

VI. Reinstatement is not an available option for practitioners with impairments caused by an irreversible medical illness or other factors not subject to rehabilitation.

## PROCEDURES:

I. #ORG# employees and other individuals who reasonably and in good faith suspect practitioner may be impaired provide a written, factual, descriptive report of the reasons for his/her suspicions to the Medical Director.

II. The Medical Director and the Department Chair of the practitioner in question discusses the incident(s) with the individual who filed the report (hereafter the "complainant").

III. Following discussion of the incident(s) complainant, if the Medical Director and Department Chairman believes there is enough information to warrant an investigation, the Medical Director convenes an investigation.

IV. Upon a determination by the Medical Director and the Department Chair that sufficient evidence exists to warrant an investigation:
   A.   Apprise #ORG#'s Administrator of the findings.
   B.   Meet with the practitioner and his or her Department Chair; and
   C.   Inform the practitioner of the results of the investigation.

V. Depending upon the severity and nature of the impairment, #ORG# may take any or all of the following actions:
   A.   Require the practitioner to enter a rehabilitation program as a condition of continued appointment and clinical privileges.
   B.   Impose appropriate restrictions or summary suspension of the practitioner's privileges if applicable.
   C.   Immediately suspend the practitioner's privileges at #ORG# or require the practitioner to discontinue their practice at #ORG# until the Medical Director and the Department Chair have concluded that rehabilitation has been completed and the practitioner has been released to return to practice.

VI. #ORG# seeks the advice of its legal counsel to determine whether any conduct is reported to law enforcement authorities or other governmental agencies, and what further steps is taken.

VII. A copy of the original report and a description of the actions taken in the practitioner's file.

VIII. The Medical Director informs the complainant follow-up action was taken, without details of such action.

IX. REHABILITATION:
   A.   The Administrator and medical staff leadership assist the practitioner in locating a suitable rehabilitation program.
   B.   The physician is not reinstated until it is established, at the discretion of the Medical Director and respective Department Chair, that practitioner has successfully completed a rehabilitation program approved by #ORG# and is capable of providing safe medical care and treatment to patients at #ORG#.

X. REINSTATEMENT:
   A.   #ORG# may consider reinstating the practitioner's job duties, privileges and/or medical staff membership, upon sufficient proof that the practitioner has successfully completed a rehabilitation program.
   B.   #ORG# obtains documentation of successful completion of the program from the physician director of the rehabilitation program where the practitioner was treated pursuant to a properly executed release of information authorization.
   C.   The documentation from the treating physician state whether:
      1.   The practitioner is currently participating or has successfully completed the rehabilitation.

2. The practitioner is in compliance with all terms of the program.
3. In the opinion of the treating physician, the practitioner is capable of resuming medical practice and providing continuous competent care of patients.

D. Upon request of #ORG# the treating physician provides information regarding the precise nature of the practitioner's condition and course of treatment.

E. If the information received indicates that the practitioner is rehabilitated and capable of resuming patient care, #ORG# implement the following precautions when restoring the practitioner's clinical privileges or allowing the practitioner to resume treating patients at #ORG#:
1. The practitioner identifies two (2) peers who are willing to assume responsibility for the care of his or her patients, in the event the practitioner is unable or unavailable to care for them; and
2. The practitioner provides #ORG# with periodic reports from his/her primary care physician, for a period of time specified by the Ad Hoc Committee.

F. Upon reinstatement, the department chair or a physician appointed by the department chair monitors the practitioner's treatment of patients and/or exercise of clinical privileges.

G. At the request of the Medical Director, the practitioner agrees to submit to an alcohol or drug-screening test prior to reinstatement and on a routine basis, if warranted.

H. All requests for information concerning the impaired practitioner are forwarded to the Medical Director.

**REFERENCES:**

**FORMS:**

**EQUIPMENT:**

**APPROVALS:**

| NAME | TITLE | DATE |
|------|-------|------|
| #APPROVER# | #APPRTITLE# | #APPRDATE# |

**Policy and Procedures**

Policy#:  #POLNUM#
Location:  #LOC#
Originating Department:  #ORIGDEPT#
Effective Date:  #EFFDATE#
Expiration Date:  #EXPDATE#

## TITLE: INFECTION CONTROL PROGRAM

### POLICY STATEMENT:

It is the policy of #ORG# to maintain an Infection Control Program.

### INTENT AND SCOPE:

The intent of this policy is to provide comprehensive guidelines that facilitate prevention of transmission of infectious diseases and compliance with regulatory guidelines. This policy affects all students, volunteers, Medical staff and employees (contract and non-contract) of #ORG#.

### DEFINITIONS:

### GENERAL INFORMATION:

   I. The Infection Control Program:
     A. Develops and implements a system for surveillance, prevention, and control of infections and communicable diseases in patients and personnel.
     B. Develops a system of identifying, reporting, investigating and decreasing the occurrence of healthcare associated infections (HAIs).
     C. Develops policies and programs that comply with regulatory agencies.

  II. The Infection Control Program is based on the needs of employees and patients served by #ORG#.

 III. Critical Components of the Infection Control Program.
     A. Surveillance Activities:
       1. Targeted infection surveillance is conducted to monitor the incidence of HAIs.
       2. Sources of data include but are not limited to:
         a. Patient rosters
         b. Microbiology reports
         c. Medical record chart
         d. Assessment findings
         e. Record of patients placed in isolation
       3. Targeted surveillance utilizes current CDC definitions for defining HAIs. Other information, such as coding data, may be used.
       4. The purpose of surveillance activities:
         a. Monitor the rate of HAIs
         b. Detect outbreaks of infection
         c. Investigate clusters and/or single cases of unusual or epidemiological significant HAIs
         d. Institute prevention and control measures
         e. Participate in quality improvement or performance improvement program

   f. Monitor infection rates for procedures with:
    1. High volume
    2. Frequent infectious complications, or
    3. High potential adverse outcomes.
   g. Monitor the effects of intervention strategies on infection rates
   h. Provide feedback to physicians, nurses, and support staff about the HAIs risk of patients.
   i. Document infections of epidemiological significance among employees
  5. Surveillance activities are directed by the Infection Control Committee and the appointed Infection Control Officer and credentialed Infection Prevention employees.
 B. Control Measures
  1. Infection Control staff is consulted during development of, and review, policies and procedures as they relate to infection control issues (examples: intravascular therapy, construction management, sterilization, disinfection and instrument decontamination.)
   a. Annual policy review is done for Exposure Control Policies (i.e., Bloodborne).
   b. All other policies are reviewed every three (3) years, and as needed if problems/issues are identified.
  2. Monitor and review epidemiologically significant infections or occupational exposures of employees.
  3. Institute control measures as indicated by surveillance activities.
 C. Regulatory Compliance
  1. Confirm compliance with regulatory agencies.
  2. Incorporate appropriate CDC recommendations for infection control related issues into policies and procedures.
  3. Conduct and/or assist with performance improvement activities as related to infection control.
  4. Reporting as required.
 D. Education
  1. Assist with identification and development of infection control topics for continuing education and training.
  2. Assist with identification and development of infection control programs for orientation classes.
  3. Provide educational programs on infection control, communicable diseases, and other related issues, yearly and as requested.

## PROCEDURES:

 I. The Quality and Safety Committee directs and supervises the Infection Control Program by:
 A. Reviewing all surveillance reports related to HAIs, communicable diseases, and other infection control issues.
 B. Reviewing the effectiveness of prevention, intervention, and control strategies in reducing HAIs rates.
 C. Reviewing the services instituted, the procedures, priorities, and/or problems identified within the past year.
 D. Provides direction for the Infection Control program and surveillance activities.

 II. The Infection Control Officer:
 A. Is appointed by the Governing Body.
 B. Serves as an infectious disease consultant.
 C. Recommends and/or supervises special studies or programs authorized by the Quality and Safety Committee.
 D. Authorizes appropriate control measures and/or studies when there is felt to be a danger to patients and/or personnel.

 III. The Infection Control Staff:
 A. Implement surveillance and other day-to-day infection control and prevention activities.
 B. Establishes surveillance objectives and conducts surveillance activities and special studies as directed.
 C. Analyzes, stratifies, and reports findings from surveillance activities and special studies.
 D. Complies with public health reporting requirements.
 E. Participates in educational programs for infection control issues, communicable diseases, and other related topics.
 F. Serves as a consultant and resource on infection control topics to all departments and services.
 G. Has the authority to implement appropriate control measures via a verbal order from the Infection Control Officer.
 H. Serves on committees pertinent to infection control issues.

**REFERENCES:**

**FORMS:**

**EQUIPMENT:**

**APPROVALS:**

| NAME | TITLE | DATE |
|------|-------|------|
| #APPROVER# | #APPRTITLE# | #APPRDATE# |

**Policy and Procedures**

Policy#: #POLNUM#
Location: #LOC#
Originating Department: #ORIGDEPT#
Effective Date: #EFFDATE#
Expiration Date: #EXPDATE#

## TITLE: INFORMED CONSENT

### POLICY STATEMENT:

It is the policy of #ORG# to provide documentation that informed consent has been granted by the patient, or a legally authorized individual to consent for the patient, prior to proposed procedures or treatments.

Specific procedures designed to protect the patient's rights are followed in the event of extreme emergency when the patient is unable to give an informed consent and no one having legal capacity to consent is immediately available, or in the event the patient is deemed to be incompetent to give informed consent.

## INTENT AND SCOPE:

This policy is intended to establish standardized requirements for obtaining consent for treatment, and meet legal and regulatory requirements.

## DEFINITIONS:

  I. Both parties validate communication—An exchange of information in which understanding is validated by both parties. This includes the elimination of technical terminology and the utilization of interpreters for both foreign and sign language when indicated.

 II. Informed Consent—An explanation given by the practitioner to the patient concerning proposed treatments and a disclosure of the risks and hazards related thereto.

III. List A Procedures—Those procedures and treatments, which require full disclosure by the practitioner or health care provider to the patient or person authorized to consent for the patient.

 IV. List B Procedures—Those procedures and treatments that do no require disclosure by the practitioner or health care provider to the patient or person authorized to consent for the patient.

  V. Emergency Treatment—Immediate medical interventions intended to stabilize/ameliorate a condition in which delay of treatment could result in death or permanent harm such as loss of limb or functional impairment.

 VI. Mark—Substitution for patient's signature when the patient is unable to sign.

VII. Minor—An individual less than eighteen (18) years of age.

VIII. Competency—An individual's capacity to understand information relative to the medical treatment decision at hand and to reason about risks and relevant alternatives against a background of stable personal values and life goals.

### GENERAL INFORMATION:

  I. Questions or problems concerning consent issues are addressed with the risk officer at the facility.

**PROCEDURES:**

I. General Consent

    A.  Upon patient entry into #ORG#, Patient Registration personnel obtain patient; parent or guardian signatures on the general consent form.

    B.  Before treatment is initiated, the nurse reviews the general consent form to determine if the form is signed. If the form is not signed, the nurse obtains the signature.

    C.  Patient rights and advance directive information is presented to the patient before a consent can be signed. For ambulatory surgery centers, patient rights and advance directive information is given to the patient prior to the day of surgery.

II. Minors

    A.  With the exception of family planning services/supplies and emergency treatment, consent from a parent or guardian is required. When the person having the power of consent cannot be contacted and actual notice to the contrary has not been given, consent may be made by one of the following:

        1.  Grandparent

        2.  Adult brother or sister (over age eighteen (18))

        3.  Adult aunt or uncle

        4.  Any adult who has care and control of the minor and written authorization from the parent.

    B.  Minors may give consent for their medical care if the minor:

        1.  Is on active duty for the Armed Forces of the United States of America.

        2.  Is sixteen (16) years of age or older and resides separate and apart from his/her parents, managing conservator, or guardian (whether with or without the consent of the parent, managing conservator, or guardian and regardless of the duration of the residence apart from such person), and is managing his/her own financial affairs regardless of the source of income.

            a.  Consents to the diagnosis of an infectious, contagious, or communicable disease which is required by law or regulation adopted pursuant to law, to be reported by the licensed physician or dentists to a local health officer, including, but not limited to venereal disease.

            b.  Is pregnant, and consent to medical and surgical treatment related to her pregnancy. Therapeutic abortions are not included. If a pregnant minor refuses to consent to reasonable necessary medical or surgical treatment related to her pregnancy, her parents may give consent. When possible, consent is best obtained from both the minor and her parent.

            c.  Consents for counseling for sexual abuse, physical abuse, or Mental Health Treatment suicide prevention.

            d.  Consents for examination and treatment for chemical addiction, chemical dependency or any other condition directly related to chemical use.

            e.  Consents to medical or surgical treatment of a contraceptive nature, except sterilizations, which cannot lawfully be performed on a minor.

            f.  Is married.

            g.  Has been emancipated by a court (minor produces a court order).

III. Emergency Treatment

    A.  If an emergency exists and the patient is not able to consent or the parent/guardian is not available, the practitioner renders emergency aid and clearly document in the medical record the following:

    B.  The nature and extent of the patient's incapacity;

    C.  The medical necessity of the proposed treatment, possible complications, etc.; and

    D.  The measures taken to locate a person or persons to function as a surrogate decision-maker.

Refer also to #ORG# Policy for Surrogate Decision Making or Guardianship of Incapacitated Patients.

IV. Oral Consent

    A.  Oral consent may be obtained when written consent cannot be obtained (i.e., when the injury precludes an otherwise competent patient from signing the form, or when the case involves a minor and the parent/guardian is contacted by phone or some other means of communication).

1. The practitioner explains the situation to the next of kin and obtains oral consent and documents this in the medical record.
2. Licensed or certified personnel confirm that oral consent was given and sign as a witness.
3. A consent form is completed and signed by the practitioner and the witness when treatment includes any procedure on List A, or when otherwise initiated by the practitioner.
   a. The form notes that oral consent was obtained by inserting the following statement over the witness signature, "The above consent form has been read and explained to _____" (individual's name).
   b. The name and relationship of the person giving oral consent is recorded on the consent form.

V. Operative Consent
   A. The practitioner determines the patient's competence, or obtains necessary consults to determine the patient's competence.
      1. The physician considers the effects of medications and/or anesthesia previously administered and the response of the individual patient. In general, consent is obtained within four (4) hours of administration of a narcotic or eight (8) hours of administration of anesthesia.
   B. The practitioner completes the form appropriately and communicates personally with the patient; disclosing all risks and alternatives, and obtains the written consent.
      1. Interpreters are used at the patient's request or when needed, for patients with hearing impairments or those who do not speak the same language as the practitioner.
      2. Family members or friends of the patient are used as interpreters only when no other interpreter is available and/or with approval of the patient.
      3. The practitioner retains full responsibility for informed consent even when an interpreter is used. The use of an interpreter is documented on the consent form, along with the name and a general description of the interpreter (i.e. Housekeeper, spouse, friend, etc.)
   C. The signing of the consent form is witnessed by licensed or certified personnel.
      1. It is not necessary for the witness to be present during the actual informed consent conference between the patient and practitioner.
      2. The witness is not required to inquire into the adequacy of the communication, but is witnessing only the signature of the patient.
   D. An operative permit is valid for fourteen (14) days.
      1. A statement on the original form may be used to document reaffirmation.
      2. Licensed or certified personnel obtain and witness the patient's signature on reaffirmation.
   E. Multiple procedures may be documented on the same consent form, if they are to be performed at the same time. A separate consent form is completed each time the patient is returned to the operative suite.
   F. The Peri-anesthesia Patient with a Do-Not-Resuscitate (DNR) Advance Directive
      1. The automatic suspension of a DNR cannot be justified for a patient requiring a surgical procedure. Patients are informed that complications encountered during the peri-anesthesia period might trigger interventions, which the patient had previously refused.
      2. The following designations are utilized for the peri-anesthesia patient based upon discussions with the patient or legal representative
      3. Full Attempt at Resuscitation: The patient or legal representative may request the full suspension of DNR status during the operative and immediate postoperative period; thereby, consenting to the use of any resuscitative measure that is appropriate based on the clinical circumstance. Notation is made in the medical record reflecting the content of this conversation. Also documented is the time at which the previously documented directive resumes.
      4. Limited Resuscitation Based on Particular Procedures
         a. The patient or legal representative may select to refuse certain resuscitative measures, such as chest compressions or defibrillation. The anesthesiologist educates the patient or legal representative about procedures that are essential to the successful use of the anesthetic agent and the proposed procedure; including, but not limited to endotracheal intubation and intravenous fluids. Notation is made in the medical record reflecting the content of this conversation. Also documented is the time at which the previously documented directive resumes.

    5. Limited Resuscitation Based on the Patient's Values and Goals

       a. The patient or legal representative may choose to trust the anesthesiologist and surgical team to use clinical judgment during the procedure, if complications are encountered. The patient's stated goals and values are used as a guide in the context of the clinical event. For example, patients may want full resuscitation procedures to manage an easily reversible event. The patient allows the physician to use clinical judgment, deferring intervention if treatment of the event is not likely to be successful or would create new and unacceptable burdens for the patient. Also document the time at which the previously documented directive resumes.

VI. Other Consent

  A. Treatment at Other Facilities

    1. The practitioner gets the consent of the patient to go to another facility for treatment.

    2. The facility performing the procedure obtains the appropriate informed consent.

  B. Specialized Consent

    Specialized consent forms are obtained for the following:

    1. Autopsy

    2. Clinical Research studies

    3. Blood

    4. Hysterectomy

    5. HIV testing

    6. Organ donation

**REFERENCES:**

**FORMS:**

**EQUIPMENT:**

**APPROVALS:**

| NAME | TITLE | DATE |
| --- | --- | --- |
| #APPROVER# | #APPRTITLE# | #APPRDATE# |

**Policy and Procedures**

Policy#: #POLNUM#
Location: #LOC#
Originating Department: #ORIGDEPT#
Effective Date: #EFFDATE#
Expiration Date: #EXPDATE#

## TITLE: INITIAL COMPETENCY AND PROVISIONAL STATUS

### POLICY STATEMENT:

It is the policy of #ORG# that the Initial Competency Evaluation is performed during a provisional period to confirm an individual Practitioner's current competence at the time initial privileges are granted, or if a currently-privileged Practitioner requests additional privileges. However, graduates of #ORG# graduate residency/fellowship training program within the past two (2) years may not be subject to this policy.

### INTENT AND SCOPE:

The purpose of this policy is that all #ORG#, state, and federal laws concerning initial Competency evaluation are in compliance. It applies to all Medical Staff and employees (contract and non-contract) of #ORG#.

### DEFINITIONS:

### GENERAL INFORMATION:

Specialties determine what constitutes most competencies for practitioners. In addition, the medical staff will review what is required such as medical knowledge and recent improvements and opportunities in care delivery. Information from the American College of Graduate Medical Education is also valued.

### COMPETENCY LEADERSHIP:

The Credentialing Committee is charged with the responsibility of monitoring compliance with this policy and procedure.
   The Medical Director and other practitioners involved with ongoing evaluation provide the Credentialing Committee with data systematically collected for evaluation that is appropriate to confirm current competence for these Practitioners during a provisional period.

### Competency Plan Development; Development of Specialty Guidelines

Competency data may be obtained by a combination of any or all of the following methods:

A. **Prospective**: Presentation of cases with planned treatment outlined for treatment concurrence, review of case documentation for treatment concurrence or completion of a written or oral examination or case simulation.
B. **Concurrent**: Direct observation of a procedure being performed or medical management through observation of Practitioner interactions with patients and staff or review of open medical records during the patient's #ORG# stay.
C. **Retrospective**: Review of case record after care has been completed. May also involve interviews of personnel directly involved in the care of the patient.

### Data Sources:

Competency data may be obtained from:

A. Personal interaction and observation with the Practitioner by the Proctor;

B. Detailed medical record review by the Proctor;

C. Structured evaluation interviews of #ORG# staff interacting with the Practitioner;

D. Surveys of #ORG# staff interacting with the Practitioner;

E. Chart audits by non-medical staff personnel based on medical staff defined criteria for initial appointees or Practitioners with newly-granted privileges; and/or

F. Data routinely obtained for evaluation as either individual case reviews or aggregate data.

## Development of the Practitioner Specific Competency Plan:

The Medical Director develops a Competency plan to accompany the recommendation to the Credentials Committee regarding the requested clinical privileges. The Competency plan defines the following:

A. Competency Methods: The Medical Director uses the specialty guidelines to determine the methods used for the Practitioner Competency Plan based on the Practitioner's background, training, reputation, demonstrated experience, and the Medical Director's first-hand knowledge of a Practitioner's current Competency.

The Medical Staff may take into account the Practitioners' previous experience in determining the approach and extent of proctoring needed to confirm current Competency. The Practitioner experience may fall into one of the following categories:

1. Recent training program graduate from another facility with less than one (1) year post training experience.

2. Practitioner with post training experience at another organization of one (1) to five (5) years.

3. Practitioner with post training experience at another organization of greater than five (5) years.

B. Competency Period and/or Patient Volume: Based on the specialty of the Practitioner, newly granted privileges are considered under Competency for either a specific period of time or for a specific number of patients/procedures. Generally, the Competency period is concluded within three to six (3-6) months. However, it may be extended for a period not to exceed a total of twentyfour (24) months from the granting of the privilege(s) that require Competency if either initial concerns are raised that require further evaluation or if there is insufficient activity during the initial period.

## Competency Site:

Unless specifically determined in a Practitioner's plan, Competency is performed onsite at #ORG# facilities. Offsite Competency may be permitted in situations where a Practitioner has skills that are needed at #ORG# on an occasional basis, where the skills and competence of the Practitioner in question are known to members of the Medical Staff of #ORG# and in situations where Practitioners are needed from local area hospitals to provide occasional coverage at #ORG#. It is up to the Medical Director to make a recommendation related to the use of off-site Competency for a specific Practitioner situation.

## PROCEDURES:

I. Responsibilities:

A. Responsibilities of the Proctor:

Proctor(s) is in good standing with the Medical Staff of #ORG# and possess the specific privileges for which the Practitioner is being evaluated.

## The Proctor:

1. Uses the proctoring methods and tools approved by the Credentials Committee, Medical Executive Committee and the BOM.

2. Assures the confidentiality of the Competency results and forms and deliver the completed Competency forms to the Medical Director.

3. Submits monthly reports and additional information requested by the Medical Director.

4. Informs the Medical Director if the Practitioner undergoing Competency is not sufficiently available or lacks sufficient cases to complete the process in the prescribed timeframe.

5. Promptly notifies the Medical Director if at any time during the Competency period, the Proctor has concerns about the Practitioner's Competency to perform specific clinical privileges or care related to a specific patient(s).

B.  Responsibilities of the Practitioner Undergoing Competency:
1.  For concurrent proctoring, make every reasonable effort to be available to the Proctor, including notifying the Proctor of each patient where care is to be evaluated in sufficient time to allow the Proctor to concurrently observe the care provided.
2.  The Practitioner secures agreement from the Proctor to attend the procedure for elective surgical or invasive procedures where direct observation is required. The required Competency is completed before the Practitioner can perform the procedure without a Proctor present.
3.  For an elective procedure where the Proctor is not available, the Practitioner may proceed with concurrence of the Proctor.
4.  In an emergency, the Practitioner may admit and treat the patient and must notify the Proctor as soon as reasonably possible.
5.  The Practitioner provides the Proctor with information about the patient's clinical history, pertinent physical findings, pertinent lab and radiology results, the planned course of treatment or management and direct delivery to the Proctor of a copy of all histories and physicals, operative reports, consultation reports and discharge summaries documented by the proctored Practitioner.
6.  The Practitioner may request from the Medical Director a change of Proctor if disagreements with the current Proctor may adversely affect his or her ability to satisfactorily complete the proctorship. The Medical Director will keep the Credentials Committee and Medical Executive Committee informed about changes in proctors.
7.  Inform the Proctor of any unusual incident(s) affecting patient care.
C.  Responsibilities of the Medical Director:
1.  Identifying the names of Practitioners in the department eligible to serve as proctors.
2.  Submitting the proctoring plan to the Credentials Committee with the recommendations for privileges;
3.  Assignment of proctors;
4.  If at any time during the proctoring period, the Proctor notifies the Medical Director that he/she has concerns about the Practitioner's Competency to perform specific clinical privileges or care related to a specific patient(s), based on the recommendations of the Proctor, the Medical Director, or designee, reviews the medical records of the patient(s) treated by the Practitioner being proctored and:
    a.  Intervenes and resolves the conflict if the Proctor and the Practitioner disagree as to what constitutes appropriate care for a patient;
    b.  Reviews the case for possible quality referrals;
    c.  Recommends to the Credentials Committee that:
        i.  Additional or revised proctoring requirements be imposed upon the Practitioner; and/or
        ii. Corrective action is undertaken pursuant to applicable corrective action procedures.
5.  At the end of the Competency period, reviews and evaluates both the case-specific reviews and aggregate Competency data relative to Medical Staff determined targets for Competency and providing the Credentialing Committee with an interpretation regarding the Practitioner's performance indicating the following:
    a.  Whether a sufficient number of cases done at #ORG# or at local hospital (i.e., via off-site Competency) have been reviewed to properly evaluate the clinical privileges requested:
        i.  If yes, the Medical Director's recommendation concerning the appointee's qualifications and competence to continue to exercise these privileges.
        ii. If no, whether in the Medical Director's opinion, the Competency period is extended for an additional period.
    b.  An evaluation and recommendation of Practitioner's clinical Competency.
D.  Responsibilities of the Quality Department Staff:
The Quality Department Staff:
1.  Audit charts being proctored as required by the proctoring plan and submit the data to the Medical Director.
E.  Responsibilities of the Credentials Committee:
The Credentials Committee:
1.  Monitors compliance with this policy and procedure, including resolving any issues or problems involved in implementing this policy;

2. Recommends to the Medical Executive Committee the proctoring plan for Practitioner for which it recommends granting of privileges;

3. Receives regular status reports related to the progress of all Practitioners required to be proctored; and

4. May make recommendations to the Medical Director regarding clinical Competency privileges based on information obtained from the proctoring process.

**REFERENCES:**

**FORMS:**

**EQUIPMENT:**

**APPROVALS:**

| NAME | TITLE | DATE |
|------|-------|------|
| #APPROVER# | #APPRTITLE# | #APPRDATE# |

**Policy and Procedures**

Policy#:  #POLNUM#
Location:  #LOC#
Originating Department:  #ORIGDEPT#
Effective Date:  #EFFDATE#
Expiration Date:  #EXPDATE#

## TITLE: ISOLATION PRECAUTIONS

### POLICY STATEMENT:

It is the policy of #ORG# that Isolation Precautions as recommended by the Centers for Disease Control and Prevention are followed and that all blood and body fluids from all patients are considered to be potentially infectious.

### INTENT AND SCOPE:

To establish safe standards of practice designed to prevent the transmission of communicable diseases. This policy applies to all Medical Staff, volunteers and employees (contract and non-contract) of #ORG# as defined by the Center for Disease Control and Prevention.

### DEFINITIONS:

I. Health Care Worker (HCW): "HCWs" refers to all the paid and unpaid persons working in health care settings. This may include, but is not limited to, physicians, nurses, aides, dental workers, technicians, workers in laboratories and morgues, emergency medical service (EMS) personnel, students, part-time personnel, temporary staff not employed by the health-care facility, and persons not involved directly in patient care but who are potentially at risk for occupational exposure (e.g., volunteer workers and dietary, housekeeping, maintenance, clerical and janitorial staff).

II. Standard Precautions (SP): Precautions that are used for contact with all patients or their blood, body fluids, secretions, excretions (except sweat), non-intact skin or mucous membranes. Standard Precautions are designed to reduce the risk of transmission of microorganisms from both recognized and unrecognized sources of infection. Standard Precautions combines the best practices of Universal Precautions and Body Substance Isolation.

III. Transmission-based Precautions: Precautions that are used for patients with documented or suspected infection with highly transmissible or epidemiologically important pathogens. These precautions are used in addition to Standard Precautions. There are three types of Transmission-based Precautions: Airborne Precautions, Droplet Precautions and Contact Precautions.

### GENERAL INFORMATION:

I. Since medical history and examination cannot reliably identify all patients infected with HIV, other bloodborne pathogens or other infectious diseases, Standard Precautions (SP) is consistently used in the care of all patients.

II. SP requires HCWs to assume that all blood and body fluids from all patients are potentially infected, and to use barriers and other protective equipment to prevent contact with all blood and body fluids.

III. SP incorporates the recommendations from the Center for Disease Control on the use of Universal Precautions for prevention of HIV and HBV Transmission in the Health Care Setting and the recommendations in OSHA's Bloodborne Standard, the Final Rule.

IV. Failure to comply with Standard Precautions is considered grounds for disciplinary action.

V. Transmission-based isolation signs are posted on patient room doors in appropriate patient care areas, as needed.

VI. To minimize the need for emergency mouth-to-mouth resuscitation, mouthpieces, resuscitation bags, and other ventilation devices are available for use in areas in which the need for resuscitation is predictable.

VII. HCWs who have exudative lesions, or weeping dermatitis, refrain from all direct patient care activities and handling patient care equipment until the condition resolves. Call Occupational Health Services or Infection Control for further information or assistance.

VIII. Pregnant HCWs are not known to be at a greater risk of contracting HIV or other diseases than non pregnant workers; however, if health care workers develop certain infections during pregnancy, the infant is at a risk of infection or health problems resulting from prenatal transmission.

## PROCEDURES:

### PROCEDURE FOR NEUTROPENIC OR IMMUNOCOMPROMISED PATIENT

I. The neutropenic or immunocompromised patient requires:
  A. Hand washing protocols,
  B. Placement in private room,
  C. No ill caregivers or visitors,
  D. No cut flowers, fresh fruits or vegetables.

### PROCEDURE FOR USING STANDARD PRECAUTIONS

I. Personal Protective Equipment (PPE) is used when in contact with all blood and body fluids from all patients. The use of barriers is based on individual judgment in determining when barriers are needed based on the individual's skills and interactions with the patient's blood, body fluid, non-intact skin, and mucous membranes.
  A. Hands are washed thoroughly prior to contact with patients, if contaminated with blood or body fluids, after gloves are removed and before gloves are put on.
  B. HCWs are trained on the proper use, selection and indications for PPE as well as the procedures for disposal or reprocessing of PPE per departmental policies.
  C. Compliance for wearing PPE is monitored within the HCWs department and is included in the HCWs performance appraisals and standards of performance.
  D. HCWs use PPE unless a rare and extraordinary circumstance occurs in which the HCW believes the use of PPE would prevent the delivery of health care or would create a risk to the worker or coworker.
    1. Such decisions not to use protective barriers in those rare and extraordinary circumstances are not applied to a particular work area or a recurring task.
    2. All such instances are documented on an Occurrence Report and investigated to determine whether prevention of similar occurrences in the future is possible.
  E. Appropriate supplies are located in each patient room or are available at a central location in each nursing unit or patient care area/department, and include protective eyewear, gowns, aprons, masks, and particulate respirators.
  F. Gloves are:
    1. Worn when touching blood and body fluids, mucous membranes, or non-intact skin of all patients and for handling items or surfaces soiled with blood or body fluids.
    2. Worn when performing venipuncture, starting IV's and for other vascular access procedures.
    3. Changed after contact with each patient.
    4. Changed after contact with contaminated site (Ex. infected wounds) and prior to contact with a clean site (IV insertion site).
  G. Particulate Respirators (PRs) are used to prevent transmission of infectious agents through the airborne route per policy for 'Tuberculosis Exposure Control Plan'.

H.   Masks and goggles or face shields are used to prevent transmission of infectious agents through procedures or situations that are likely to generate droplets or splashing of blood or body fluids.
    1.   When worn, masks are used only once and discarded in an appropriate receptacle. EXCEPTION; A mask that is a particulate respirator may be reused for the duration of one (1) shift of work.
    2.   Masks are not lowered around the neck and re-used.
    3.   When worn, all masks are to cover the nose and mouth.
I.   Disposable gowns or aprons are worn during procedures that are likely to cause splattering or splashes of blood or body fluids. These items prevent clothing and/or skin contamination with blood or body fluids.
J.   Disposable gowns or aprons are worn and discarded per departmental policy. Any protective apparel is appropriately discarded or laundered when visibly contaminated with blood or body fluids.

## STANDARD PRECAUTIONS: SHARPS

Precautions are taken to prevent injuries caused by needles, scalpels, and other sharp instruments or devices during procedures, when cleaning used instruments, during disposal of used needles and when handling sharp instruments after procedures.

  I.   After use, disposable syringes and needles, scalpel blades, and other sharp items are placed in puncture-resistant containers for disposal, containers are located as close as practical to the area.

 II.   To prevent needle stick injuries, contaminated needles are not recapped, purposely bent or broken by hand, removed from disposable syringes, or otherwise manipulated by hand. Needle cutting devices are not used.

III.   When situations or procedures occur where recapping is unavoidable; a mechanical recapping device or a safe one-handed recapping technique is used. Following are examples of such occurrences:
    A.   On a Psychiatric unit or in a correctional facility where sharp containers are not readily available because of patient safety issue.
    B.   Some blood gas sampling equipment necessitates recapping of the needle.
    C.   Administration of incremental doses of medication to the same patient.

IV.   One-handed recapping procedure:
    A.   Lay the needle cap on a flat surface.
    B.   Guide the contaminated needle into the cap using one hand only.
    C.   Secure needle cap by exerting pressure against a solid surface, such as a bedside table.

 V.   Nursing personnel check the sharps containers in their patient care areas at the beginning of their shift and notify Environmental Services if they are at full line or have full indicator displayed.

VI.   Environmental Services routinely check the sharps containers and replace when at full line or have full indicator displayed. Sharps containers are not allowed to overflow and are securely closed when removed for transport. Environmental Services transports closed containers in an upright position to the biohazard disposal area.

VII.   Reusable sharps:
    A.   Reusable sharps that are contaminated with blood or other potentially infectious materials are not stored or processed in a manner that requires HCWs to reach by hand into the containers where these sharps are placed.
    B.   Contaminated REUSABLE sharps are placed in a container until properly reprocessed. The containers are puncture resistant, labeled or color-coded and leak proof on sides and bottom.

## STANDARD PRECAUTIONS: SPECIMENS

  I.   All patient specimens are considered potentially infectious and are handled and transported in accordance with Infection Control Policy for Laboratory Services.

 II.   All specimen containers are handled wearing gloves.

III. All patient specimens are placed in a sturdy leak proof container with a secure lid to prevent leakage during transport. Care is taken when collecting and handling specimens to avoid contamination of the outside of the container and or the laboratory requisition accompanying the specimen. All specimens require the use of secondary containers with a biohazard label.

IV. All specimens are transported to the Laboratory in a carrier container.

## STANDARD PRECAUTIONS: LINEN

I. All used and soiled linen is considered contaminated and potentially infectious and is handled accordingly.

II. All used linen is placed in an appropriate bag in the patient's room, or in the area where it was used, and then transported to the soiled linen holding area. Wet linen is placed in an impervious bag.

III. Laundry is handled as little as possible with a minimum of agitation.

IV. HCWs who have contact with contaminated laundry wear appropriate PPE as per departmental policy.

## STANDARD PRECAUTIONS: WOUND DRESSINGS

I. All wound dressings are disposed of in a manner that confines and contains any blood or body fluids that may be present:
  A. Small dressings can be enclosed in the disposable glove used to remove the dressing. Grasp the dressing in the palm of the glove and pull the glove off inside out containing the dressing inside the glove. These items are disposed of in the regular trash container.
  B. Larger dressings are removed using gloved hands and placed in an impervious bag, securely sealed, and placed in the regular trash. If the dressing is dripping it is discarded in a red bag.

## STANDARD PRECAUTIONS: SPECIAL WASTE

I. Definitions (Refer to #ORG# Policy on 'Waste Disposal')
  A. Bulk blood and body fluids (Per CDC Universal Precautions Guidelines)—aggregate volume of 100 ml or more; saturated items—those that drip without compression.
  B. Microbiological waste—cultures, stocks, discarded vaccine vials, etc.
  C. Pathological waste—severed body parts, tissue, products of conception, etc.
  D. Sharps—hypodermic needles, scalpel blades, broken contaminated glass, etc.

II. All special waste is contained and discarded in appropriate red plastic trash bag or sharps container and is considered potentially infectious.

III. Special waste is transported by Environmental Services staff to the storage area for final processing and/or transport.

IV. Ordinary waste, such as paper supplies, boxes, IV tubing & cannulas, dressings, etc, is discarded in the regular trash. Ordinary waste is considered dirty and is handled appropriately.

## STANDARD PRECAUTIONS: DECONTAMINATION

I. All used medical device(s) and/or equipment is considered potentially infectious and handled as contaminated.

II. Medical devices or instruments that require sterilization or disinfection are thoroughly cleaned before being exposed to the disinfectant or sterilant. All disinfectants or sterilants must be EPA registered and approved by Infection Control. The manufacturer's instruction for the use of the disinfectant or sterilant is followed.

III. Appropriate gloves are worn for all decontamination procedures.

IV. Non-disposable articles and/or equipment contaminated with blood or body fluids are bagged or placed in a leak proof container before sending for decontamination and reprocessing.

V. Contact lenses used in trial fittings are disinfected after each fitting by using a hydrogen peroxide contact lens disinfecting system or if compatible, with heat (78°C–80°C [172.4°F–176.0°F]) for ten minutes.

VI. HIV is inactivated rapidly after being exposed to commonly use chemical germicides at concentrations much lower than those used in routine practice. Acceptable solutions include:
   A.   Commercially available—Infection Control approved and EPA registered—chemical germicides.

## STANDARD PRECAUTIONS: REPROCESSING INSTRUMENTATION AND MEDICAL EQUIPMENT:

HCWs comply with current guidelines for disinfection and sterilization of reusable devices used in invasive procedures. As part of standard infection control practice, instruments and other reusable equipment used in performing invasive procedures must be appropriately decontaminated, disinfected and sterilized as follows:

I. All items are decontaminated prior to disinfection or sterilization. Decontamination removes all visible organic material.

II. Critical Items: Equipment and devices that enter the patient's vascular system or other normally sterile areas of the body are sterilized before being used for each patient.

III. Semi Critical Items: Equipment and devices that touch intact mucous membranes but do not penetrate the patient's body surfaces undergo high-level disinfection before being used for each patient. These items may be sterilized before being used.

IV. Non Critical Items: Equipment and devices that do not touch the patient or that only touch intact skin of the patient are cleaned and disinfected with approved cleaning agent or as indicated by the manufacturer.

## STANDARD PRECAUTIONS: CLEANING AND DECONTAMINATION OF BLOOD/BODY FLUIDS SPILLS

I. All blood and body fluid spills are cleaned up using SP and a facility approved chemical germicide or a solution of sodium hypochloride (1:10 dilution of household bleach) or by using a Blood Spill Kit as approved by the organization.

II. Additional PPE (such as goggles or tongs) are used if splattering of contaminated materials or contact with sharps is anticipated during the clean up procedure.

III. For large spills of infectious agents:
   A.   Wear gloves, eye protection, and disposable gown/aprons.
   B.   Absorb as much of the spill with an absorbent material or paper towel and place in a biohazard bag. Collect any sharp objects with forceps or other mechanical services and place in a sharps container.
   C.   Using an approved chemical germicide solution, clean the spill site of all visible blood in a circular pattern from the outer edges to the center to avoid spreading contamination.
   D.   Spray the site with approved chemical germicide and allow to air dry for fifteen (15) minutes.
   E.   Discard all disposable material used to decontaminate the spill and any contaminated personal protective equipment into a biohazard bag.
   F.   Wash your hand with soap and water.

## TRANSMISSION-BASED ISOLATION PRECAUTIONS

I. Certain disease processes require additional measures beyond SP to prevent transmission to others. The CDC has published Transmission-based Isolation Guidelines that include a listing of diseases and appropriate precautions.

II. Patient Placement: Patient placement is a significant component of isolation precautions.
   A.   A private room is indicated when the patient has poor hygienic habits, contaminates the environment, or cannot be expected to assist in maintaining infection control precautions to limit transmission of infectious materials (examples: infants, children, draining wounds, and patients with altered mental status.)

B. When possible, patients with highly transmissible or epidemiologically important microorganisms are placed in a private room. Example:
  1. Patients with TB, Measles, or Chicken pox are placed in negative air pressure rooms.
  2. Patients with multi resistant organisms are placed in private rooms.
  3. Patients with draining wounds that are not adequately contained with a dressing.
  4. Patients may require Droplet and Contact Precautions based on the site on colonization or infection
C. When a private room is not available, an infected patient is placed with an appropriate roommate.
D. Infection Control staff is consulted for any concerns regarding patient placement.

III. Implementing Transmission-Based Isolation Precautions:
  A. Obtain an order from the physician.
    1. In the event that the practitioner fails to take appropriate action, the Manager/Charge nurse has the authority to implement Transmission Based Precautions until Infection Control staff can be reached.
    2. Infection Control clinical staff can write a verbal order for Isolation from the Infection Control Officer.
  B. A STOP SIGN is posted on the patient's door (Mark the type of precaution needed. DO NOT write the patient's disease or any information that gives information about the diagnosis on the sign or any other item that is easily viewed by family, visitors or others.)
  C. Document on the Nursing assessment/flow sheet or nursing notes the type of precautions needed (Airborne, Droplet, or Contact).
  D. Enter appropriate precaution in the comment section when using the computer to enter an order request.
  E. Necessary supplies (i.e. PPE's, surgical masks, PR's etc.) are available.
  F. Educate the patient/family the reason for isolation precautions and document that they verbalize understanding.

IV. Transporting Patients Transmission-Based Precaution:
  A. Limit the movement or transport of Airborne Precaution patients or those infected with virulent or epidemiologically significant organisms to essential purposes only.
  B. When transportation is necessary, it is important that appropriate barrier (e.g. surgical masks, impervious dressings) are worn or used by the patient to control the spread of infection.
  C. Personnel in the area to which the patient is taken are notified of the impending arrival of the patient and the appropriate precautions to be used to reduce the risk of transmission of infection.
  D. The patient's medical record is available for pick up by other departments if the patient mustleave his/her room. The patient's medical record accompanies the patient to all ancillary departments.
  E. Patients are informed of ways that they can assist in preventing the transmission of their infection to others.

V. Accessing the Isolation Room:
  A. Enter the anteroom, when present or the room wash hands thoroughly and put on the appropriate precaution attire (i.e. gown, mask, or particulate respirator.)
  B. When finished in the Isolation room, discard appropriate items in the trash container provided in the room. Wash hands thoroughly before leaving. For Airborne precaution, exit through the anteroom.

VI. Discontinuing Airborne Precautions:
  A. Airborne precautions remain in effect until discontinued by order of the physician.
  B. Inform the patient they have been removed from Airborne Precaution.
  C. Remove or cover the STOP SIGN on the patient's door.

## PROCEDURE FOR AIRBORNE PRECAUTION

I. In addition to SP, patients with a disease that is transmitted by the airborne route are placed in Airborne Precaution. Three diseases are known to be transmitted by the airborne route: pulmonary or laryngeal Tuberculosis (TB) measles and chicken pox (or disseminated varicella zoster).

II. Patients with known or suspected airborne diseases are placed in a negative air pressure room with a sign on the door instructing all who enter to wear appropriate respiratory protection. Refer to #ORG# Policy on 'Tuberculosis Exposure Control Plan'. Airborne Isolation rooms, with negative air pressure, are available on specific Floors and in the Emergency Department.

III. Psychiatric patients requiring Airborne Precaution are transferred to a negative air pressure room, following consultation with Infection Control staff.

IV. Any patient diagnosed with varicella (Chicken pox) or non-immune with a history of exposure to Chicken pox within the past twentyone (21) days is placed in Airborne Precaution. The patient is assigned only to personnel known to be immune to varicella. Chicken pox patients are considered infectious until all lesions are scabbed over.

V. To help maintain negative pressure, the hallway and the anteroom doors remain closed at all times. Enter and exit the patient's room from the anteroom, not from the hallway door. When it is necessary to use the hallway door (i.e., to bring in equipment) close door immediately after use.

VI. Visitors for a patient in Airborne Precaution are limited and have appropriate precautions explained to them and it is expected that the visitors observe those precautions.

## DROPLET PRECAUTIONS:

I. In addition to SP, patients with diseases that are transmitted by the droplet route are placed in Droplet Precaution.

II. Transmission by droplet occurs when the patient is infected with organisms that are large particle droplets that can be generated during coughing, sneezing, talking, or performance of procedures.

III. When a private room is not available and cohorting is not achievable, maintain spatial separation of at least three (3) feet between the infected patient and other patients and visitors.

IV. Special air handling and ventilation are not necessary and the door may remain open.

V. Staff who are working within three (3) feet of the patient wear a surgical mask or may put on mask when entering the room if it is anticipated that they may have close contact with the patient.

VI. Patients with MRO's may require Droplet precaution based on the site of colonization and/or infection.

## CONTACT PRECAUTIONS:

I. In addition to SP, patients who have or are suspected to have infection or colonization with epidemiologically significant organisms that spread by the contact route are placed in Contact Precaution. (See attachment 1 for detailed information). Refer to #ORG# Policy on 'Management of Multiple Resistant Organisms'.

II. Gloves are worn for all activities that require touching the patient (including dry skin) and/or environmental surfaces and patient care equipment in the patient's environment. Gloves are changed after contact with infective material (fecal material and wound drainage). Remove gloves before leaving the patient's environment and wash hands immediately with an antimicrobial agent. After glove removal and hand washing, confirm that hands do not touch potentially contaminated environmental surfaces or items in the patient's room in order to avoid transfer of infection to other patients.

III. Gowns are worn if substantial contact with the patient, environmental surfaces, or items in the patient's environment is anticipated. Remove gown before leaving the patient's environment. After gown removal, confirm that clothing does not contact potentially contaminated environmental surfaces in order to avoid transfer of infection to other patients.

IV. A private room may be indicated. Consult with Infection Control staff for assistance with determining appropriate patient placement.

V. Use of non-critical patient care equipment is dedicated to the patient infected, or colonized, with multiple resistant organisms. (See policy on 'Management of Multiple Resistant Organisms')

**REFERENCES:**

**FORMS:**

**EQUIPMENT:**

**APPROVALS:**

| NAME | TITLE | DATE |
|------|-------|------|
| #APPROVER# | #APPRTITLE# | #APPRDATE# |

**Policy and Procedures**

Policy#: #POLNUM#
Location: #LOC#
Originating Department: #ORIGDEPT#
Effective Date: #EFFDATE#
Expiration Date: #EXPDATE#

## TITLE: LABORATORY SPECIMEN LABELING REQUIREMENTS

### POLICY STATEMENT:

It is the policy of #ORG# that all specimens submitted to the Laboratory for analysis and evaluation are labeled such that patient safety is confirmed and in accordance with accreditation and regulatory standards established by the Clinical Laboratory Improvement Amendment of 1988 and the College of American Pathologists.

### INTENT AND SCOPE:

This policy is intended to communicate labeling requirements for laboratory specimens and is applicable to all Medical Staff and employees (contract and non-contract) of #ORG# responsible for collecting, managing, submitting, delivering, and analyzing specimens to and/or by Laboratory Services.

### DEFINITIONS:

### GENERAL INFORMATION:

Unique specimen labeling criteria require separate policies:

   I. Sexually transmitted disease testing

  II. Tissue specimens

 III. Cytology specimens

 IV. Blood Bank Specimens

### PROCEDURES:

  I. In the presence of the patient, all specimens submitted to the Laboratory are legibly labeled with ALL of the following requisite information:
    A. The patient's full name, correctly spelled, and date of birth;
    B. The patient's unique numeric identifier as medical record number, financial number, etc.;
    C. The date and time of specimen collection; and
    D. The collector's initials.

  II. All specimens are accompanied by an appropriate electronic or paper request from a licensed practitioner. Without regard to priority, the order includes ALL the following requisite information:
    A. The name of the practitioner ordering the test;
    B. The name of the requested procedure;
    C. The patient's first and last name, correctly spelled;
    D. The patient's date of birth;

E. The patient's unique numeric identifier;
F. The patient care area from which the request originated;
G. The specimen source (if other than blood);
H. The date and time of collection; and
I. The collector's initials.

III. Labels are affixed to the specimen container. A specimen is NOT acceptable if the label is only on the lid of the container or the secondary transport device (biohazard bag).

IV. Slides submitted for examination are labeled in pencil on the frosted end of the slide. Slide-specimens are NOT acceptable if identifying data are only on the slide folder. Slides are labeled with ALL of the following requisite information:
A. The patient's full name, correctly spelled;
B. The patient's date of birth;
C. The patient's medical record number; and
D. The date of specimen collection.

V. Improperly labeled specimens are discarded. Labeling issues for specimens such as spinal fluid, amniocentesis, or tissue samples are addressed through a discussion between the Technologist or Histologist, Pathologist, and practitioner who obtained the specimen. Final decisions regarding labeling acceptability are made by the Pathologist. If corrective re-labeling is permitted, it is performed by the practitioner who obtained the specimen, and a chartable comment documenting the event is placed into the computer report record by the reporting Technologist.

VI. If a specimen is rejected, the patient care area from which the sample was submitted is notified, and notification is documented as an order footnote in the Laboratory Information System. If notification of the practitioner is deemed necessary, it is the responsibility of personnel from the patient care area that ordered the test to do so.

**REFERENCES:**

**FORMS:**

**EQUIPMENT:**

**APPROVALS:**

| NAME | TITLE | DATE |
|---|---|---|
| #APPROVER# | #APPRTITLE# | #APPRDATE# |

**Policy and Procedures**

Policy#:  #POLNUM#
Location:  #LOC#
Originating Department:  #ORIGDEPT#
Effective Date:  #EFFDATE#
Expiration Date:  #EXPDATE#

## TITLE: LICENSURE, REGISTRATION, AND CERTIFICATION REQUIREMENTS

**POLICY STATEMENT:**

It is the policy of #ORG# that all employees in jobs that require a license, registration, and/or certification, have evidence of current authorization to practice his/her profession/vocation. The Managers coordinate the ongoing verification of required licensure, registration, and/or certification for employees and are responsible for contract/agency personnel. The Medical Director is responsible for the ongoing credentialing of practitioners practicing in #ORG# 's facilities.

**INTENT AND SCOPE:**

This policy is intended to establish a standardized mechanism for verification of licensure or certificate of employees, and to meet regulatory requirements. This policy applies to all #ORG# employees (contract and non-contract) in jobs that require licensure, registration, and/or certification.

**DEFINITIONS:**

**GENERAL INFORMATION:**

  I. It is the employee's responsibility to observe his/her renewal date and comply prior to expiration.

  II. All managers monitor continued compliance for all current #ORG# employees of licensure, registration or certification requirements by verifying renewal of current licensure, registration or certification documents with State regulating agencies.

  III. It is the employee's responsibility to immediately notify his/her department head regarding any changes/restrictions in his/her license, registration and/or certification.

  IV. Employees are suspended if their current license, registration or certification is not renewed and documentation is not provided by the end of the expiration date.

  V. Suspended employees are terminated thirty (30) days after the expiration date for failure to update licensure, registration, and/or certification.

  VI. Contract/Agency personnel who fail to present the required written documentation to the appropriate personnel on or prior to the expiration date are restricted from work at #ORG#.

**PROCEDURES:**

  I. New Hires
    A.   The applicant for jobs that require a license, registration or certification, notes the license number and expiration date, if any, on the employment application.

B. The recruiter verifies, using the internet access to the appropriate State regulating agencies, the current licensure, registration or certification of the applicant prior to the applicant's employment start date, or

C. The applicant presents the original document to the recruiter prior to his/her employment start date

II. Verification of Ongoing Authorization to Practice

A. Current licensure, registration, and/or certification information as stated in General Information Section II of this policy is maintained in the HR file.

B. In the event an employee fails to renew his/her license, registration, and/or certification by the expiration date, the employee goes to the Human Resources Department in person to provide his/her original license, registration, and/or certification renewal from the appropriate State Board.

C. Employees failing to renew his/her license, registration, and/or certification by the expiration date are not permitted to work that require license, registration, and/or certification.

D. Employees not in compliance with this procedure are:

1. Suspended until such time as evidence of current license, registration, or certification is provided up to a maximum of thirty (30) days. Termination follow if such evidence is not presented to the department head prior to the expiration date.

    a. The employee's supervisor prepares the notice of suspension using the appropriate counseling form

    b. The employee's supervisor presents the counseling form to the employee

2. Terminated (with administrative approval) if at anytime their license, registration, or certification is denied or revoked.

III. Contract/Agency Personnel

A. Contract/agency personnel in jobs that require license, registration, or certifications are ultimately responsible to provide written documentation of renewal on, or prior to, the expiration date to appropriate personnel in the department where they are assigned to work. The department director/manager is responsible for maintaining these records.

B. It is the responsibility of the supervisor and/or designated staff to verify the original license when the contract/agency employee reports to work at #ORG#

C. The Department director/manager prepares a list of all contract/agency personnel with current licensure, registration and certification on a monthly basis and forwards it to Human Resources.

D. Verification of licensure, including license number and expiration date, is kept on file in the department/division or central staffing office where contract/agency personnel are assigned to work.

E. Contract/Agency personnel not providing the proof of current licensure, registration and or certification are not be permitted to work at #ORG#

**REFERENCES:**

**FORMS:**

**EQUIPMENT:**

**APPROVALS:**

| NAME | TITLE | DATE |
|------|-------|------|
| #APPROVER# | #APPRTITLE# | #APPRDATE# |

**Policy and Procedures**

Policy#:  #POLNUM#
Location:  #LOC#
Originating Department:  #ORIGDEPT#
Effective Date:  #EFFDATE#
Expiration Date:  #EXPDATE#

## TITLE: LIFE SAFETY MANAGEMENT PLAN

### POLICY STATEMENT:

It is the policy of #ORG# to establish, support and maintain a Life Safety Management Program to protect patients, visitors, personnel and property from fire and products of combustion.

### INTENT AND SCOPE:

This policy is intended to cover all aspects of fire safety and life safety including fire plans, training, drills, inspection, testing and maintenance of suppression, detection, package and portable systems, as well as continuous compliance with 2000 LS Code including but not limited to the Statement of Conditions (SOC) and shall conform to NFPA 101 and NFPA 99 codes as appropriate. It all applies to all visitors, students, Medical Staff and employees (contract and non-contract) of #ORG#.

### DEFINITIONS:

### GENERAL INFORMATION:

    I. Administration. The Administration of #ORG# accepts responsibility for the life safety program's leadership, program effectiveness, and continuous program review and improvement.

    II. Facility Safety as part of the Quality and Safety Committee. The Life Safety Management Plan is administered by Facility Safety who has the authority from the Administration to oversee all aspects of the program.

    III. Department Managers. The department managers are responsible for Life Safety within their departments. They set the example and inspire a focus in those they supervise by assuring the Life Safety policies and procedures are carried out.

    IV. Employees. Employees must cooperate with all aspects of the Life Safety Plan and maintain a competent level of understanding as to their roles and responsibilities within the plan. They are also required to complete annual mandatory education programs and have a firm understand of the Life Safety Plans in the departments in which they work.

    V. Physicians and Volunteers. Physicians and Volunteers are responsible to cooperate with all aspects of the Life Safety Plan and maintain a competent level of understanding as to their rolls and responsibilities within the plan.

    VI. PROGRAM COMPONENTS:
        A.  Protect patients, personnel, visitors, and property from fire and other products of combustion.
           This is accomplished through the fire protection systems and orientation and training of staff.
           When planning for the size, configuration, and equipping the space renovated, altered or new construction,
              #ORG# has the architect design new or renovated spaces based on:

Guidelines for Design and Construction of Hospitals and Health Care Facilities, 2001 ed. (per CMS & NJDOH codes). All applicable state rules and regulations.Similar standards or guidelines if applicable.

B. Inspect, test, and maintain fire protection and life safety systems, equipment, and components on a regular basis.

1. The facility has a fire alarm system that is inspected, tested, and maintained. That upon activation (1) minimizes smoke transmission through control of designated fans and/or dampers in air handling and smoke management systems and (2) transmits the fire alarm to the local fire department. 25% are done quarterly for a 100% inspection once a year. During the annual inspections, all of the devices are inventoried and tested for proper operation.

2. Smoke detectors, pull stations, panel alarms and each zone is tested for proper operation. Any system failure or malfunction is immediately addressed. Documentation of these inspections and follow-up maintenance work is kept on file in administrator. This contractor also and inspects, tests and maintains all automatic fire extinguishing systems. Each water flow and tamper switch is tested as required by the contractor. The fire pumps have flow tests as required or as requested by the insurance carrier and are cycled at least weekly.

3. Operation and inspections are performed on the fire protection systems by Engineering and by outside expert contractors. Inspections are performed and filed in administration. The inspection is part of the Preventive Maintenance system and signed by the person performing the inspection. The system is connected to a central monitoring station. If the fire alarm is activated, it is immediately transmitted to the fire department through the central monitoring station.

4. Inspect, test, and maintain all automatic fire-extinguishing systems.

All automatic fire extinguishing systems are inspected, tested, and maintained by Engineering and by outside expert contractors. Records of these inspections are maintained in administration.

Manage portable fire extinguishers, including monthly inspection, regular maintenance, and guidelines for their identification, placement, and use.

Fire extinguishers in the corridors and patient care areas are mostlythe "ABC" type unit. Fire extinguishers are located in marked areas throughout the facility. Staff is trained to operate the fire extinguishers in orientation and at least annually. Fire extinguishers are visually inspected monthly by Engineering. Actions taken are on file in administration. All fire extinguishers are inspected at least annually by outside expert contractors.

C. Report and investigate Life Safety Code (LSC) and fire protection deficiencies, failures, and user errors.

Life safety code deficiencies may be identified through scheduled hazard surveillance surveys, fire drills, or through observation while performing other duties. These should be reported to Plant Operations immediately. The Facility Engineering Manager is responsible for reporting all fire protection system deficiencies, failures, or user errors to the Quality and Safety Committee.

D. Review proposed purchase of bedding, drapes/curtains, furniture, decorations, wastebaskets, and other equipment for fire safety.

E. Emergency procedures that address:

1. Facility-wide fire-response needs

2. Area-specific needs and fire evacuation routes,

3. Specific staff responsibilities when fire discovered,

4. Specific staff responsibilities when alarm sounds,

5. Specific responsibilities in building evacuation.

## Emergency procedures for response to fire are addressed in the facility Fire Plan.

F. Specific roles and responsibilities of personnel, physicians, and other licensed independent practitioners (LIPs) at the fire's point of origin are addressed in department-specific fire plans and the facility's Fire Plan.

G. Specific roles and responsibilities of personnel, physicians, and other LIPs away from a fire's point of origin are addressed in department-specific fire plans and the facility's Fire Plan.

H. Specific roles and responsibilities of other personnel who participate in the fire plan, such as volunteers, students, and physicians are addressed in department-specific fire plans and the facility's Fire Plan.

I. Use and functioning of fire alarm systems.

## Fire Protection Systems Description:

J. Fire Alarm Central Station

K.  Alarms System
    1.  Description and Zones
    2.  Smoke detectors
    3.  Manual Pull Stations
L.  Sprinkler System
M.  Wet Chemical System in the Dietary Department
N.  Fire Extinguishers
O.  Building Safety Features

Description: The organization provides life safety features which conform to applicable NFPA 101 code sections relating to existing, Corridors, compartmentation, and other pertinent features.

Smoke Compartments: Compartmentation is a conscious focus for all construction.

Compartmentation features of floor-to-structure corridor wall and space enclosures are present. Smoke walls are shown on documents located in administration.

Smoke Doors: Located in smoke partitions. They are installed per NFPA requirements and have closures and positive latching hardware.

Magnetic Holders: Installed on all smoke doors when the door may impede normal flow of routine traffic. The holders are controlled by the Fire Alarm System, which are release the door upon alarm.

Elevator Recall: In the event of an alarm situation, the elevators are controlled manually by personnel from the fire brigade.

Exits: All exits are marked with signage conforming to NFPA 101 codes.

P.  Specific roles and responsibilities in preparing for building evacuation

**Specific roles and responsibilities for each department in preparing for building evacuation are addressed in the facility-wide Evacuation Plan.**

Q.  Location and proper use of equipment for evacuating or transporting patients to areas of refuge. Staff is trained on location of wheelchairs and stretchers to be used in evacuating or transporting patients to safe areas of refuge.

R.  Building compartmentalization procedures for containing smoke and fire. Staff is trained on compartmentalization structures, such as smoke and fire doors, and horizontal and vertical evacuation procedures to be used in the event of a fire.

S.  A policy defines the processes required for Interim Life Safety Measures. Interim Life Safety Measures are implemented based upon established criteria. The administrative actions are selected and implemented based on the project. Inspections are conducted and documented daily.

T.  Orientation/education program that addresses:
    1.  Facility fire evacuation routes.
    2.  Specific role at the fire's point of origin.
    3.  Specific role away from the fire's origin.
    4.  Use and function of the fire alarm systems.
    5.  Specific role in preparing for building evacuation.
    6.  Location and use of equipment for evacuating, or
    7.  Transporting patients in a fire.
    8.  Building compartmentalization procedures for containing smoke and fire.

**All employees are provided education regarding life safety management at the facility and departmental level. It is accomplished in the following manner:**

U.  All new employees are provided with facility life safety management education during Orientation. They are also oriented to departmental life safety management responsibilities during their initial three months of employment by the Department Manager.

V.  All employees receive at least annual life safety management education.

W.  The departmental programs are based on employee needs, either requested or assessed, and are coordinated by the respective department managers.

X.  Ongoing monitoring of performance regarding actual or potential risks related to one or more of the following:

**Performance improvement standards for Emergency Management is monitored on an ongoing basis and reported to the Quality and Safety Committee and include at least one of the following:**

1. Staff knowledge and skills
2. Level of staff participation
3. Monitoring and inspection activities
4. Emergency and incident reporting
5. Inspection, preventive maintenance, and testing of equipment.

## ANNUAL EVALUATION

The Quality and Safety Committee annually reviews the Life Safety Management Plan's objectives, scope, performance and effectiveness in meeting current codes, standards, and the needs of #ORG#.

**PROCEDURES:**

**REFERENCES:**

**FORMS:**

**EQUIPMENT:**

**APPROVALS:**

| NAME | TITLE | DATE |
| --- | --- | --- |
| #APPROVER# | #APPRTITLE# | #APPRDATE# |

**Policy and Procedures**

Policy#:  #POLNUM#
Location:  #LOC#
Originating Department:  #ORIGDEPT#
Effective Date:  #EFFDATE#
Expiration Date:  #EXPDATE#

## TITLE: MANAGEMENT OF DISRUPTIVE INDIVIDUALS

### POLICY STATEMENT:

It is the policy of #ORG# to adopt and enforce reasonably implemented "zero tolerance" of disruptive individuals or activity that could disrupt and/or adversely affect patient treatment, facility operations or other patients, visitors and/or employees.

### INTENT AND SCOPE:

The intent of this policy is to communicate an appropriate process to manage threatened and actual disruptive behavior. This policy applies to all patients, visitors, students, Medical staff and employees (contract and non-contract) of #ORG#.

### DEFINITIONS:

### GENERAL INFORMATION:

Disruptive individuals create apprehension for patients, visitors and healthcare providers.

### PROCEDURES:

  I. When an individual threatens disruptive behavior or becomes disruptive and/or disorderly, the following countermeasures are initiated:

    A.  The individual is asked to calm him or herself.

    B.  #ORG# Police or Security Department (#ORG# PD) is notified and reasonable attempts made by officers to calm the disruptive individual. If the disruptive behavior continues, the individual is escorted from the campus or arrested for disorderly conduct, as appropriate.

    C.  If cooperation is refused after reasonable attempts have been made to escort the disruptive individual from the campus, #ORG# PD may take the disruptive individual into temporary detention to allow evaluation of the danger potential.

    D.  An attempt is made to determine if the individual is an imminent threat to themselves or others and an application for detention for mental evaluation may be warranted by #ORG# PD.

    E.  Individuals on #ORG# campus or off-site clinics are asked to remove or cover provocative attire that may incite violence. Such attire includes, but is not necessarily limited to, insignia and "colors." Individuals are requested to refrain from hand signs or signals that may incite violence. If requests are refused, the individual(s) are escorted from the premises.

    F.  Any graffiti, tagging or marking is prohibited on #ORG# property. Such markings are reported to Plant Operations or Environmental Services for immediate removal, repair or coverage. Any individual(s) found to be engaging in this form of criminal mischief are taken into custody by #ORG# PD and prosecuted.

  II. When information is received through a reliable sources of the impending or actual arrival of a victim of violence who may be at continued risk of additional violence while a patient, the supervisor or designee initiates the following countermeasures:

    A.  Contact and advise #ORG# PD through the police communications office of the possibility of continued violence against the patient.

    B.  List the arriving victim of violence under the pseudonym "John or Jane Doe" and inform all appropriate personnel and care units of the patient's status as a "no information" patient and a victim at continued risk of violence.

    C.  Prepare for potential media involvement.

III.  The Supervisor or designee of the assigned patient care area:

    A.  Places the patient in a center hallway room if possible.

    B.  Places the patient in a private room whenever possible.

    C.  Issues the patient an assumed name to be listed on the hallway identification area.

IV.  A pediatric area is utilized for victims who are under age eighteen (18), provided beds are available.

V.  Rival combatants are located on separate floors whenever possible.

Only immediate family or legal guardians may visit victims at continued risk of violence under the age of eighteen (18).

**REFERENCES:**

**FORMS:**

**EQUIPMENT:**

**APPROVALS:**

| NAME | TITLE | DATE |
|------|-------|------|
| #APPROVER# | #APPRTITLE# | #APPRDATE# |

**Policy and Procedures**

|  |  |
| --- | --- |
| Policy#: | #POLNUM# |
| Location: | #LOC# |
| Originating Department: | #ORIGDEPT# |
| Effective Date: | #EFFDATE# |
| Expiration Date: | #EXPDATE# |

## TITLE: MEDICAL EQUIPMENT MANAGEMENT

### POLICY STATEMENT:

It is the policy of #ORG# that the scope of Medical Equipment Management Plan provides for the repair, preventive maintenance, cost containment, code compliant, and safety inspections of equipment.

### INTENT AND SCOPE:

#ORG#'s Medical Equipment Management Plan is designed to assess and control the physical and clinical risks of all equipment used in the diagnosis, treatment, and monitoring of our patients. This policy applies to all Medical Staff and employees (contract and non-contract) of #ORG#.

### OBJECTIVES:

1. Identify and evaluate medical equipment prior to use; assessing function, risks, and maintenance requirements and to maintain equipment incident history and assessments.
2. Maintain accurate equipment inventory for the entire facility.
3. Provide appropriate preventive inspection, testing, and maintenance all the while assessing and minimizing risk.
4. Prepare and implement policies and procedures that govern acquisition, inspection, and maintenance of all equipment used for patients.
5. Act on reports of equipment hazard notice and alerts.
6. Ensure compliance with the Safe Medical Devices Act.
7. Report and investigate equipment problems, failures, abuse and errors.
8. Ensure that all users of the equipment are trained, addressing equipment capabilities, use, limitations of equipment, operating instructions, safety protocols, emergency interventions, reporting of problems, user errors and failures.
9. Competency assessments of users prior to use and annually thereafter.
10. Quality control analysis.
11. User compliance monitoring.

### DEFINITIONS:

SMDA: Safe Medical Device Act.

## GENERAL INFORMATION:

I. #ORG#'s Medical Equipment Management Plan includes the following:
A. Minimizes the clinical and physical risks of equipment through inspection, testing and regular maintenance;
B. Establishes criteria for identifying, evaluating and inventorying equipment which is included in the program;
C. Provides education to personnel on the capabilities, limitations and special applications of equipment; operating, safety and emergency procedures of equipment; the procedures to follow when reporting equipment management problems, failures and user errors; and the skills and/or information to perform maintenance activities.

## PROCEDURES:

The Manager of Biomedical Engineering is responsible for maintaining the Medical Equipment Management Program. Each department director of #ORG# is responsible for orienting new staff members to the capabilities, limitations, special applications of equipment, basic operating and safety procedures, emergency procedures if failure occurs, maintenance responsibilities, if applicable, and the reporting procedures for equipment problems, failures and user errors.

I. The selection and acquisition of medical equipment:
   A. A "Capital-Equipment Purchase/Lease Req." form is completed by each department for replacement or new equipment. Engineering determines if the equipment meets appropriate space requirements, load and phase requirements, Laboratory requirements, minimum safety standards of three (3) wire AC line cord with health grade plug, appropriate warranties and manufacturer's reliability prior to purchase. If the equipment does not meet the above specifications, it may not be ordered and an alternate choice may be submitted for approval.

II. Establishing criteria for identifying, evaluating and taking inventory of medical equipment to be included in the Equipment Management Program:
   A. All mechanical and electrical patient care equipment is evaluated prior to use, based on function including diagnosis, care, treatment and monitoring; physical risks associated with use, maintenance requirements and history of equipment incidents. All incoming and existing equipment meeting the evaluation criteria are included in the equipment management program.
   B. All new equipment is inventoried and inspected prior to use for patient care use. Equipment that fails electrical safety tests is not approved for use until the deficiencies have been corrected. There is a current inventory of all equipment included in the equipment management program.
   C. See Biomedical Equipment Management Policies: Equipment Management Numbers Formula, Equipment Management Inventory Policy, Equipment Management Inventory Form, Additions to Equipment Management Inventory Form and Deletions from Equipment Management Inventory Form.

III. Assessing and minimizing clinical and physical risks of equipment through inspection, testing and maintenance:
   A. All mechanical and electrical patient care equipment is evaluated prior to use. Semiannual preventive maintenance and safety inspections are completed on all equipment with risk scores of seven or higher. The results of inspections and maintenance are kept in the equipment management program.
   B. Incident history is documented and maintained in the equipment management program and in the Biomedical Engineering office/administration. Equipment displaying unusual repair history or unusual incidence of injury to staff or patients is evaluated for necessary changes/replacement.
   C. All other clinical electrically powered equipment is safety inspected annually by the Biomedical Engineering and a tag or sticker is affixed.
   D. The Biomedical Engineering Supervisor develops preventive maintenance procedure for all medical devices in the facility. The preventive maintenance procedures are developed using the manufacturer's preventive maintenance recommendations, National Fire Protection Agency (NFPA) standards and ECRI standards.

IV. Hazard notices and recalls:
   A. All product safety alerts, hazard notices and recalls are directed to the Biomedical Engineering Supervisor. In the event the notices are not directed to the Biomedical Engineering Supervisor. The Biomedical Engineering Supervisor checks the clinical equipment inventory to screen for equipment matches and evaluates the severity of the risk. In most cases, the notices may be addressed without removing equipment from service. In the event equipment is removed from service, the equipment is replaced with a safe effective substitute. Biomedical Engineering impounds equipment removed from use due to recall notices until it can be rendered safe.
   B. The Biomedical Engineering Supervisor reports monthly to the Quality and Safety Committee on any hazard notices and recalls affecting #ORG# and all follow up activities undertaken.

V. Monitoring and reporting of medical device incidents resulting in death, serious injury or serious illness of any individual as per Safe Medical Device Act Of 1990:

A. The Safe Medical Device Act of 1990 requires that device user facilities (including hospitals, outpatient diagnostic and treatment facilities, nursing homes, ambulatory surgical facilities) report incidents to the device manufacturer when the facility determines a device has or may have caused or contributed to the death or serious injury of an individual. The facility must also send a copy of the report to the Food and Drug Administration (FDA) in the case of a death.

B. #ORG# has established methods for reporting these events:

C. The appropriate personnel are notified immediately.

D. All packaging and disposable materials are returned.

E. The device is inspected and control settings and any damage is recorded. The equipment is bagged, tagged and sequestered by Biomedical Engineering or Risk Management.

F. An investigation is conducted.

G. The Risk Manager is responsible for managing the Safe Medical Device Act reporting process.

VI. Investigation and reporting of equipment management problems, failures and user errors:

A. All equipment failures and user errors are investigated and reported monthly to the Quality and Safety committee. Included in the report is the total number of equipment malfunctions, abuse, operator error, and Preventive Maintenance PM) repairs. The report also lists all departments that meet or exceed the thresholds set by the safety committee. In the event the equipment problem was caused by operator error, the user(s) are in-serviced on the operation and use of the equipment.

VII. The Medical Equipment Management Plan includes a Medical Equipment Orientation and Education Program:

A. Thorough training is provided regarding the capabilities, limitations, special applications of equipment, basic operating and safety procedures, emergency procedures if failure occurs, maintenance responsibilities, if applicable, and the reporting procedures for equipment problems, failures and user errors included in the program by department directors or designees in involved departments. All users/maintainers of equipment are tested for competency according to the components of their job specifications.

VIII. Performance Standards:

A. There is a planned, systematic, interdisciplinary approach to process design and performance measurement, analysis and improvement related to organization wide safety. The organizational Quality and Safety Committee develops and establishes performance measures and related outcomes, in a collaborative fashion, based on those priority issues known to be associated with the healthcare environment. Performance measures and outcomes are prioritized based upon high risk; high volume, problem prone situations and potential or actual sentinel event related occurrences. Criteria for performance improvement measurement and outcome indicator selection is based on the following:

1. The measure can identify the events it was intended to identify:
   a. The measure has defined data elements and allowable values;
   b. The measure can detect changes in performance over time;
   c. The measure allows for comparison over time within the organization or between the organization and other entities;
   d. The data intended for collection are available;
   e. Results can be reported in a way that is useful to the organization and other interested stakeholders.

2. The Quality and Safety Committee on an ongoing basis monitors performance regarding actual or potential risk related to one or more of the following:
   a. Staff knowledge and skills;
   b. Level of staff participation;
   c. Monitoring and inspection activities;
   d. Emergency and incident reporting;
   e. Inspection, preventive maintenance and testing of safety equipment.

B. Other performance measures and outcomes are established by the Quality and Safety Committee, based on the criterion listed above. Data sources, frequency of data collection, individual(s) responsible for data collection, aggregation and reporting is determined by the Quality and Safety Committee.

C.  If the Quality and Safety Committee feel a team approach (other than the Safety Committee) is necessary for performance and process improvement to occur, the Committee follows the organization's performance improvement guidelines for improvement team member selection. Determination of team necessity is based on those priority issues listed (high risk, volume and problem prone situations and sentinel event occurrence). The Committee reviews the necessity of team development, requesting team participation only in those instances where it is felt the Committee's contributions toward improvement is limited (due to specialty, limited scope and/or knowledge of the subject matter). If team development is deemed necessary, primarily, team members are selected on the basis of their knowledge of the subject identified for improvement, and those individuals who are "closest" to the subject identified. The team is interdisciplinary, as appropriate to the subject to be improved.

D.  Performance improvement monitoring and outcome activities are presented to the Quality and Safety Committee by the Manager of Biomedical Engineering at least on a quarterly basis, with a report of performance outcome forwarded to the Medical Executive Committee and Governing Body quarterly.

E.  The following performance measures are recommended:
1.  Percent of staff able to demonstrate their knowledge and skill of their role and expected participation in the medical equipment management plan;
2.  Percent of performance assessments/evaluations reflecting competence to provide service;
3.  Number of equipment incidents reported;
4.  Percent of PMs completed on time.

IX. Emergency Procedures:
A.  Equipment, which meets #ORG# criteria for critical to patient safety, have emergency procedures in the event a malfunction or failure occurs. Equipment considered critical to patient safety includes life support, life sustaining or other critical equipment whose malfunction or failure may result in an adverse patient outcome.

B.  Each department develops and follow specific clinical response procedures in the event of an equipment failure:

C.  Equipment is removed from service and tagged immediately.

D.  Institute clinical emergency procedures required ensuring patient care is not compromised.

E.  If replacement equipment is necessary, an equipment rental company (depending on the kind of equipment) is notified to obtain a replacement.

F.  Engineering, Risk Manager and Performance Improvement are notified of the failure.

G.  An incident report is completed describing the failure.

X. Annual evaluation of the Medical Equipment Management Plan:
A.  The annual evaluation of the Medical Equipment Management Program includes a review of the scope according to regulatory and accreditation guidelines. A comparison of the expectations and actual results of the program is evaluated to determine if the goals and objectives of the program were met. The overall performance of the program is reviewed by evaluating the results of performance improvement outcomes. The overall effectiveness of the program is evaluated by determining the degree that expectations where met.

B.  The performance and effectiveness of the Medical Equipment Management Program are reviewed by the Quality and Safety Committee and Administration.

C.  See Annual Evaluation of the Effectiveness of the Medical Equipment Management Program.

**References:**

**Forms:**

**Equipment:**

**APPROVALS:**

| NAME | TITLE | DATE |
|------|-------|------|
| #APPROVER# | #APPRTITLE# | #APPRDATE# |

**Policy and Procedures**

Policy#:  #POLNUM#
Location:  #LOC#
Originating Department:  #ORIGDEPT#
Effective Date:  #EFFDATE#
Expiration Date:  #EXPDATE#

## TITLE: MEDICAL PLANS OF CARE

### POLICY STATEMENT:

It is the policy of #ORG# that the Medical Staff develops implements and evaluates medical plans of care that are clinically relevant and based upon scientifically recognized authorities. All Medical Plans of Care are approved by the Medical Staff and are intended to improve the quality and utilization of care for patients.

### INTENT AND SCOPE:

This policy is intended to provide guidance for the development, review, approval and monitoring of Medical Plans of Care at #ORG#. Medical Plans of Care are used as guidelines. Medical Plans of Care may be modified to address documented patient needs.

Medical Plans of Care are represented in several formats and include plans of care, clinical pathways, preprinted routine orders, and protocols.

This policy applies to all medical staff and employees (contract and non-contract) of #ORG#.

### DEFINITIONS:

I. Plans of Care—An educational tool that is developed to assist practitioners in clinical decision making. These include:
   A. Guidelines, which are accepted principles for patient management.
   B. Guidelines which are recommendations for patient management identifying a particular strategy;
   C. Practice policies; or
   D. Practice options.

II. Clinical Pathways or Best Practice Models—A standardized interdisciplinary approach to a specific medical condition. Clinical pathways must include:
   A. Interdisciplinary, collaborative care plans
   B. Physician order, protocols and/or guidelines for treatment
   C. Sequencing of interventions
   D. Comprehensive aspects of care
   E. Expected patient outcomes.

III. Preprinted Routine Orders—A set of physician orders, which document a course of treatment for a specific set of patients. Although orders are preprinted, initiation requires the written order by a physician and all preprinted orders should be reviewed and modified as needed by the physician for individual patient needs.

IV. Protocols—A predefined series of activities related to a specific set of patients or procedures. A qualified practitioner may initiate the protocol without physician orders provided the defined conditions are met.

**GENERAL INFORMATION:**

I.  All Medical Plans of Care must have documented the following prior to approval process
   A.  Indicated need for development
   B.  Identification of process improvement with base line measurement improvement opportunities
   C.  Implementation process and time line.
   D.  Identification of any barriers to the implementation process
   E.  Research and or standard used for basis of development
   F.  Summary of development process
   G.  In addition all Clinical Pathways or Best Practice Models must have:
      1.  Goals of pathway including measurable patient outcomes
      2.  Tools for measurement of outcomes
      3.  Process for outcome measurement
      4.  Reporting structure of improvement

**PROCEDURES:**

I.  A team, committee or department identifies a need for of a Medical Plan of Care.

II.  The request is documented by contacting the Manager.

III.  The Manager facilitates the referring source in documenting the needed data to determine, the validity and priority of the request and then forwards all documentation to the Quality and Safety Committee for review.

IV.  The Quality and Safety Committee under the leadership of the Chair, who is also the Medical Director, determines the validity and priority of the plan. The committee assists in identifying physician and nurse champion(s) and medical department for development of the Medical Plan of Care.

V.  The Manager facilitates the development of the Medical Plan of Care by:
   A.  Facilitating the project including coordinating the nursing care elements and the medical care elements.
   B.  Providing education on the definitions of the process and needs for the development of the plan of care.
   C.  Providing guidance during the approval process.
   D.  Coordinating the quarterly reporting of the development, implementation, evaluation, and outcome measurement.

VI.  Completed Medical Plans of Care are reviewed by the physician champion(s) and medical department assigned and the nursing champion(s). Once the plan has been approved for content, it is forwarded to the Quality and Safety Committee for approval.

VII.  The Quality and Safety Committee is responsible for reviewing the format, outcomes measurement, standard evidence basis of the medical plan, content and approval of the plan.

VIII.  Upon approval by the Quality and Safety Committee the Medical Plan of Care is forwarded to the Medical Executive Committee for consideration and final approval.

Upon approval by the Medical Executive Committee the plan can be implemented.

**REFERENCES:**

**FORMS:**

**EQUIPMENT:**

**APPROVALS:**

| NAME | TITLE | DATE |
|------|-------|------|
| #APPROVER# | #APPRTITLE# | #APPRDATE# |

**Policy and Procedures**

Policy#:  #POLNUM#
Location:  #LOC#
Originating Department:  #ORIGDEPT#
Effective Date:  #EFFDATE#
Expiration Date:  #EXPDATE#

## TITLE: MEDICAL STAFF CODE OF CONDUCT

### POLICY STATEMENT:

It is the policy of #ORG# that all Medical Staff members practicing at #ORG# treat others with respect, courtesy, and dignity and conduct themselves in a professional and cooperative manner.

### INTENT AND SCOPE:

It is the intent of this policy to secure optimum patient care by promoting a safe and cooperative health care environment and to prevent or eliminate (to the extent possible) conduct that:

  I. Disrupts the operation of #ORG#;

  II. Affects the ability of others to do their jobs;

  III. Creates a "hostile work environment" for #ORG# employees or other medical staff members;

  IV. Interferes with an individual's ability to practice competently; or

  V. Adversely affects or impacts the community's confidence in #ORG#'s ability to provide quality patient care.

This policy is also intended to address sexual harassment of employees, patients, other members of the Medical Staff, and others, which is not tolerated.

In dealing with all incidents of inappropriate conduct, the protection of patients, employees, physicians, and others at #ORG# and the orderly operation of the Medical Staff and #ORG# are paramount concerns. Complying with the law and providing an environment free from sexual harassment are also critical. This policy applies to all Medical Staff (contract and non-contract) of the #ORG#.

### DEFINITIONS:

  I. Inappropriate conduct—behavior that includes, but is not limited to, the following:
    A. Inappropriate physical contact with another individual that is threatening or intimidating;
    B. Degrading or demeaning comments regarding patients or their families, nurses, physicians, #ORG# personnel or #ORG#;
    C. Profanity or similarly offensive language while at #ORG# and/or while speaking with patients, nurses or other #ORG# personnel;
    D. Inappropriate medical record entries concerning the quality of care being provided by #ORG# or any other individual or otherwise critical of #ORG#, other Medical Staff members or personnel;
    E. Derogatory comments about the quality of care being provided by #ORG#, another Medical Staff member, or any other individual or otherwise critical of #ORG#, another Medical Staff member, or any other individual that are made outside of appropriate Medical Staff and/or administrative channels;

F. Threatening or abusive language directed at patients, nurses, #ORG# personnel, or other physicians (e.g., belittling, berating, and/or threatening another individual;

G. Refusal to abide by medical staff requirements as delineated in the Medical Staff Bylaws and any amendments thereto, Credentialing Manual, and Rules and Regulations (including, but not limited to, emergency call issues, response times, medical record keeping, and other patient care responsibility, failure to participate on assigned committees, and an unwillingness to work cooperatively and harmoniously with other members of the Medical Staff and #ORG# staff); and/or;

H. "Sexual Harassment," which is defined as any verbal and/or physical conduct of a sexual nature that is unwelcome and offensive to those individuals who are subjected to it or who witness it. Examples include, but are not limited to, the following:

1. Verbal: innuendos, epithets, derogatory slurs, off-color jokes; propositions, graphic commentaries, threats, and/or suggestive or insulting sounds;

2. Visual/Non-verbal: derogatory posters, cartoons, or drawings; suggestive objects or pictures; leering; and/or obscene gestures;

3. Physical: unwanted physical contact, including touching, interference with an individual's normal work movement, and/or assault; and

4. Other: making or threatening retaliation as a result of an individual's negative response to harassing conduct or reporting harassing conduct.

Investigative Team—A group of individuals assigned to investigate potential violations of this policy as appointed by the Medical Director and Administrator.

## GENERAL INFORMATION:

I. Issues of the inappropriate conduct of an employee are dealt with in accordance with #ORG# Policies. Issues of conduct by members of the Medical Staff (hereinafter referred to as "practitioners") are addressed in accordance with this policy.

II. This policy outlines steps that can be taken in an attempt to resolve complaints about inappropriate conduct exhibited by practitioners. However, there may be a single incident of inappropriate conduct, or a continuation of conduct, that is so unacceptable as to make such steps inappropriate and that requires immediate disciplinary action. Therefore, nothing in this policy precludes an immediate referral to the Medical Staff Peer Review Committee or the elimination of any particular step in the policy when dealing with a complaint about inappropriate conduct.

## PROCEDURES:

## PROCEDURE WHEN A CONCERN IS RAISED

I. Nurses and other #ORG# employees who observe, or are subjected to, inappropriate conduct by a practitioner notify their supervisor about the incident or, if their supervisor's behavior is an issue, they notify the Medical Director or Administrator. Any practitioner who observes such behavior by another practitioner notifies the Medical Director or Administrator immediately. Upon learning of the occurrence of an incident of inappropriate conduct or potential inappropriate conduct, the person to whom the incident is reported requests that the individual document it in writing. In the alternative, the person to whom the incident is reported may document the incident as reported.

A. Documentation to include the following:

1. The date and time of the questionable behavior;

2. The name of any patient or patient's family member who may have been involved in the incident, including any patient or family member who may have witnessed the incident;

3. The circumstances which precipitated the situation;

4. A factual and objective description of the questionable behavior;

5. The names of other witnesses to the incident, if any;

6. The consequences, if any, of the behavior as it relates to patient care, personnel or facility operations;

7. Any action taken to intervene in or remedy the situation, including the date, time, place, action, and name(s) of those intervening; and

8. The name and signature of the individual reporting the complaint of inappropriate conduct.

B. The person to whom the incident is reported forwards the report to the Investigative Team.

C. The Investigative Team reviews the report and may meet with the complainant or the individual who prepared it and/or any witnesses to the incident to ascertain the details of the incident.

## PROCEDURE WHEN INAPPROPRIATE CONDUCT HAS LIKELY OCCURRED

I. If the Investigative Team determine that an incident of inappropriate conduct has likely occurred, it proceeds as follows:

A. The Investigative Team meets with the practitioner and advises him or her of the following:

1. That any attempt to confront, intimidate or retaliate against the individual(s) who reported the behavior in question, whether the reporting individual's identity is disclosed or not, is a violation of this policy and grounds for further immediate disciplinary action; and

2. That such conduct is inappropriate and is not tolerated.

B. The Investigative Team counsel and educate the practitioner during such meeting about the concerns and the necessity to modify the behavior in question.

C. A copy of this policy is provided to the practitioner during the meeting.

D. The Investigative Team documents in writing the meeting with the practitioner.

II. If the Investigative Team prepares any documentation for a practitioner's file regarding its efforts to address concerns with the practitioner, the practitioner is apprised of that documentation and is given an opportunity to respond in writing. Any such response is then kept in the practitioner's confidential file along with the original report and the Investigative Team's documentation.

## PROCEDURE WHEN ADDITIONAL COMPLAINTS ARE MADE

I. If additional complaints are received concerning a practitioner, including complaints regarding retaliation for making the initial complaint, the Investigative Team may have an additional meeting with the practitioner (following the steps in Paragraph I above) as long as it believes that there is still a reasonable likelihood that those efforts resolve the concerns. If a meeting is to take place, the practitioner is informed that if the inappropriate conduct continues, Medical Director and Administrator are informed. Further, the practitioner is also informed at the meeting that failure to abide by the terms of this policy is grounds for summary suspension of his or her Medical Staff membership and clinical privileges.

II. The meeting with the practitioner is documented in writing by the Investigative Team, with a follow-up letter to the practitioner. The letter documents the content of the discussion and any specific action the practitioner has agreed to perform. The documentation and letter is placed in the practitioner's file.

III. The practitioner is apprised of the documentation and is given an opportunity to respond in writing. Any such response is then kept in the practitioner's confidential file along with the report and all documentation by the Investigative Team.

IV. At any point in this process, however, the Investigative Team may refer the matter to the Medical Staff Peer Review Committee for review and action in lieu of conducting another meeting with the practitioner. When it makes such a referral, the Investigative Team may also suggest a recommended course of action for the practitioner (e.g., behavior modification course, development of conditions for continued practice for the individual, suspension).

## PROCEDURE WHEN INAPPROPRIATE CONDUCT CONTINUES

I. If after the second meeting with the Investigative Team, the offending behavior continues, the Medical Director and Administrator are immediately fully apprised of the situation as well as the previous warning issued to the practitioner and the actions that were taken to address the concerns. The Medical Director and Administrator meet with the practitioner about the inappropriate conduct. During such meeting inform the individual that a single recurrence of the offending behavior result referral of the matter to the Medical Staff Peer Review and Medical Executive Committees to be formally investigated pursuant to the Bylaws of the Medical Staff.

II. If after the final warning the offending behavior recurs, the matter is referred to the Medical Staff Peer Review Committee to be formally investigated pursuant to the Bylaws of the Medical Staff.

III. Should the Peer Review Committee's investigation result in an action by the Medical Staff Executive Committee that entitles the practitioner to request a hearing under the Fair Hearing Plan, the practitioner is provided with copies of all relevant complaints so that he or she can prepare for the hearing.

In order to effectuate the objectives of this policy, and except as otherwise may be determined by the Medical Director and Administrator.

**REFERENCES:**

**FORMS:**

**EQUIPMENT:**

**APPROVALS:**

| NAME | TITLE | DATE |
|------|-------|------|
| #APPROVER# | #APPRTITLE# | #APPRDATE# |

**Policy and Procedures**

Policy#:  #POLNUM#
Location:  #LOC#
Originating Department:  #ORIGDEPT#
Effective Date:  #EFFDATE#
Expiration Date:  #EXPDATE#

## TITLE: MEDICAL STAFF CREDENTIALING PROCESS

### POLICY STATEMENT:

It is the policy of #ORG# that credentialing and privileging of practitioners resides within the Medical Staff Office in close collaboration with #ORG# physician leadership and the Board of Managers (BOM).

### INTENT AND SCOPE:

Establish and implement a standard for the credentialing and privileging process of practitioners who provide care in #ORG#. This policy applies to all Medical staff and allied health professionals (employee, contract and non-contract) of #ORG#.

### DEFINITIONS:

The following definitions apply to the provisions of the credentialing and privileging process. The definitions are presented in alphabetical order.

 I. Allied Health Professional (AHP)—an individual other than a licensed physician, dentist, oral surgeon or podiatrist who is qualified by academic and clinical training and by prior and continuing experience and current competence in a discipline, which the Board of Managers has determined to allow to practice at #ORG#. An "Independent AHP" is an individual licensed by the state of #STATE# and permitted by #ORG# to provide Services at #ORG# without the direction of immediate supervision of a Medical Staff Member. A "Dependent AHP" is an individual who functions in a medical support role to and under the direction and supervision of a Medical Staff Member. Allied Health professionals are members of the Medical Staff.

 II. Board of Managers or Board (BOM)—the governing body of #ORG#. As appropriate to the context and consistent with the Bylaws of #ORG# and delegations of authority made by the Board, it may also mean any committee of the Board or any individual authorized by the Board to act on its behalf on certain matters.

 III. Chief Executive Officer (CEO)/Administrator—the individual appointed by the Board of Managers as the Chief Executive Officer/Administrator of #ORG# to be responsible for the overall executive supervision and management of #ORG#. The CEO may, consistent with his responsibilities under the Bylaws of #ORG#, designate a representative to perform his responsibilities under the Medical Staff Bylaws and related Manuals.

 IV. Clinical Privileges (Privileges)—the permission granted pursuant to the Bylaws to a practitioner to provide services.

 V. Credentialing Committee (credentialing committee)—Receives and reviews the qualification of each applicant for appointment or reappointment or modification of appointment for clinical privileges, including practitioners on the medical staff and allied health professionals. Members reviews at two (2) year intervals the Bylaws, Rules and Regulations and medical staff policies and procedures.

 VI. Dentist—an individual with a D.D.S. or D.M.D. degree, who is licensed to practice dentistry and whose practice is in the area of general dentistry or a specialty thereof.

VII. #ORG#—to include the facility located at #ADDRESS# #STATE# as well as other entities owned and operated by #ORG#.

VIII. Medical Staff (Staff)—the organizational component of #ORG# that includes all credentialed practitioners.

IX. Medical Staff Bylaws (Bylaws)—the Bylaws of the Medical Staff of #ORG# and its related documents. Related documents are any one or more of the following documents as appropriate to the context:
 A. Medical Staff Credentialing Policy and Procedure
 B. Fair Hearing Plan
 C. Medical Staff Rules and Regulations
 D. Medical Staff Peer Review Policy

X. Medical Staff Executive Committee (MEC)—the committee that receives and acts upon reports and recommendations from committees, departments and officers of the medical staff.

XI. Medical Staff Member in Good Standing or Member in Good Standing—a practitioner who has been appointed to the Medical Staff and who is not under either a full or partial suspension.

XII. Medical Staff Year—the 12-month period commencing on January 1st.

XIII. Oral Surgeon—an individual with a D.D.S. or D.M.D. who is licensed to practice dentistry in the State of #STATE# and who has successfully completed an approved postgraduate program in oral surgery.

XIV. Physician—an individual with a M.D. or D.O. degree, who is licensed to practice medicine in the State of #STATE#.

XV. Podiatrist—an individual with a D.P.M. degree, who is licensed to practice podiatry in the State of #STATE#.

XVI. Practitioner—unless otherwise expressly provided, any physician, oral surgeon, dentist, podiatrist, psychologist or allied heath practitioner who either: (a) is applying for appointment and clinical privileges; or (b) currently holds appointment to the Medical or Dental Staff and has specific delineated clinical privileges.

XVII. Prerogative—a participatory right granted, by virtue of Staff category or otherwise, to a staff member or Allied Health Professional.

XVIII. Special Notice—written notification sent by certified mail, return receipt requested, or by personal delivery service with signed acknowledgement or receipt.

XIX. Temporary Member—a practitioner who has been granted temporary privileges for a time period to be determined on a case-by-case basis due to patient care need, but not to exceed one hundred twenty (120) days.


**GENERAL INFORMATION:**

**PROCEDURES:**

 I. APPOINTMENT PROCEDURES AND TERM OF APPOINTMENT
 A. Request For Application:
 1. A request for application is submitted to the medical staff office and includes full name, office address and telephone number. Upon receipt of a request, MSO forwards to the applicant the following documents:
 a. Cover letter explaining the Pre-Application for Membership.
 b. Pre-Application for Membership document
 2. Upon return the Pre-Application for Membership is reviewed by MSO and forwarded to the medical director for review.

3. If the medical director concludes the applicant is ineligible for appointment under said qualifications, the CEO informs the applicant by special notice through MSO and the application does not proceed through the credentialing process.

4. The CEO informs the applicant by special notice if the applicant is denied an application because he/she is seeking only clinical privileges for particular services that either is not provided at #ORG# or is provided pursuant to a closed Staff or exclusive contract policy.

5. An individual whose application has been denied is not entitled to the procedural due process rights provided in the Fair Hearing Plan and the denial is not considered an adverse action reportable under the Health Care Quality Improvement Act and/or relevant state law.

6. Unless the application is denied pursuant to the above process, MSO accepts the application and proceed with the credentialing process.

7. Following approval of the pre-application, MSO staff forwards the following documents to the applicant:
   a. Credentialing Application.
   b. Privileges specific to applicant's specialty.
   c. Medical Staff Bylaws, Rules and Regulations and the Fair Hearing Plan.
   d. Miscellaneous Forms (i.e., Pharmacy Signature Form, and Medicare/Medicaid Acknowledgement Forms).

8. The practitioner completes the Credentialing Application in full and presents the signed application and clinical privileges request to MSO by mail or in person.

B. Effect of Application
   1. The applicant signs the application and in so doing;
      a. Attests to the correctness and completeness of all information furnished and acknowledges that any misstatement or misrepresentation in or omission from the application, whether intentional or not, constitutes grounds for denial of said application or revocation of membership and privileges;
      b. Signifies his/her willingness to appear for interviews in connection with his/her application;
      c. Agrees to abide by the terms of the Bylaws, Rules and Regulations, Department and #ORG# policies if granted appointment and/or clinical privileges, and to abide by the terms thereof in all matters relating to consideration of the application without regard to whether or not appointment and/or privileges are granted;
      d. Agrees to maintain an ethical practice and to provide continuous care to his/her patients;
      e. Agrees to notify MSO, promptly and in writing, of any change made or proposed in the status of his professional license or permit to practice;
   2. DEA or state controlled substance registration;
      a. Professional liability insurance coverage, and membership/employment status;
      b. Clinical privileges at other institutions/facilities/organizations, including but not limited to, revocation of privileges, summary suspension, probation, monitoring, proctoring;
      c. The issuance of a reprimand and/or any other disciplinary or corrective action and on the status of current or the initiation of new malpractice claims.
      d. Authorizes and consents to #ORG# representatives consulting with prior associates or others, including malpractice insurance carriers, who may have information bearing on professional or ethical qualifications and competence and consents to their releasing and #ORG# inspecting all records and documents, including but not limited to documents generated in the peer review process that may relate to the applicant's character, qualifications and competence;
      e. Releases from any liability all those who, in good faith and without malice, review, act on or provide information or documentation described in XVI or other information or documentation regarding the applicant's background, experience, clinical competence, professional ethics, utilization practice patterns, character, health status, and other qualifications for staff appointment and clinical privileges.

C. Processing the Application
   1. Applicant's Burden and Proof of Identity;
      a. The applicant has the burden of producing adequate information for a proper evaluation of his/her experience, training, current competence, utilization practice patterns, ability to work cooperatively with others, character and health status.
      b. The applicant also has the burden of resolving any doubts about these or any of the qualifications required for Staff appointment or the requested Staff category, Department, or clinical privileges,

and satisfying any reasonable requests for information or clarification made by appropriate Staff or #ORG# authorities.

c. In addition to other requests made of the applicant a final affirmative decision on Staff appointment and/or on the granting of all or particular privileges may be conditioned upon the applicant obtaining a physical and/or psychiatric evaluation, at the applicant's expense, with appropriate tests by a physician(s) acceptable to the MEC.

d. If the applicant is to obtain a physical and/or psychiatric evaluation, the CEO sends him/her a special notice so indicating and the deadline for the response, which is not later than thirty (30) days from the date of the notice.

e. Failure to obtain the required evaluation within that period is deemed a withdrawal of the application, unless the Credentialing Committee determines the failure to respond was caused by circumstances beyond the applicant's control. The CEO sends the applicant special notice of any deemed withdrawal.

f. An applicant who has had his/her application withdrawn due to failure to comply with this provision is not entitled to the procedural due process rights provided in the Fair Hearing Plan and the withdrawal is not itself considered an adverse action reportable under the Health Care Quality Improvement Act and/or relevant state law.

D. Verification Information;

1. The completed application is submitted to MSO. MSO staff organizes and coordinates collection and primary source verification of the references, licensure and other qualification evidence submitted or required, and promptly notifies the applicant of any gaps in or any problems in obtaining the required information.

2. A special notice is sent indicating information not received the nature of the additional information the applicant is to provide and the time frame for response, which does not exceed thirty (30) days from the date of the notice.

3. Failure to respond in a satisfactory manner by that date is deemed a withdrawal of the application, unless the credentialing committee determines the failure to respond was caused by circumstances beyond the control of the applicant.

4. The CEO sends the applicant special notice of any deemed withdrawal. An applicant who has their application withdrawn due to failure to comply with this provision is not entitled to the procedural due process rights provided in the Fair Hearing Plan and the withdrawal is not in itself considered an adverse action reportable under the Health Care Quality Improvement Act and/or relevant state law.

5. Verification by MSO includes, without limitation:

   a. Reasonable effort to confirm with the primary source information contained on the application;

   b. Send a list of clinical privileges requested by the applicant to his training programs for the initial application only.

   c. Requesting other specific information and ratings, as appropriate, on aspects of the applicant's performance at affiliation which may bear on their qualifications for Staff appointment/reappointment or the privileges requested, including;

      i. basic medical knowledge and professional judgment;

      ii. clinical competence and technical skill;

      iii. cooperativeness and ability to work with others;

      iv. medical record currency; and

      v. Availability to meet patient care needs.

   d. Requesting information from applicable sources, including governmental agencies and National Practitioner Data Bank, as is required under state or federal law.

   e. Upon verification and receipt of required information staff forwards the file to the chair of each department the applicant seeks privileges and request evaluation of the credentials file.

   f. A completed credentials file is when all documents requested have been received and all information verified by MSO staff.

E. Department Evaluation:

1. The chair of each department, the applicant seeks privileges or the applicable department committee reviews the application and its supporting documents.

2. The Department Chair, or their designee, interviews the applicant.

   3. If a Department Chair or their designee requires further information, they may defer transmitting their report for no more than thirty (30) days except for good cause.
   4. In this instance, the applicable Department Chair or designee notifies MSO office, and the credentialing committee Chair.
   5. A special notice is sent to the applicant and states if the applicant is to provide additional information or a specific release/authorization to allow #ORG# representatives to obtain information or documentation within the deadline for the response which is not to exceed thirty (30) days.
   6. Failure to respond by the deadline is deemed a withdrawal of the application unless the credentialing committee determines the failure to respond was caused by circumstances beyond the applicant's control.
   7. The CEO sends the applicant special notice of any deemed withdrawal. An applicant whose application is withdrawn due to failure to comply with this provision is not entitled to the procedural due process rights provided in the Fair Hearing Plan and the withdrawal is not itself considered an adverse action reportable under the Health Care Quality Improvement Act and/or relevant state law.
   8. MSO forwards credentialing information to the credentialing committee following approval of the applicant by the Department Chair (s) which is indicated by the signature of the Chair.
F. Credentialing Committee Evaluation:
   1. The credentialing committee thoroughly reviews the information provided by the MSO staff in reference to the applicant. The application, supporting documentation, reports from the Department Chair or the designee, and any other relevant information is available for members review.
   2. The credentialing committee may, at its discretion, interview the applicant or designate one or more of its members to do so.
   3. The credentialing committee prepares a written report including a summary of interview conducted.
   4. The credentialing committee may defer transmitting its report to MEC for no longer than forty-five (45) days if the credentialing committee required further information.
   5. The credentialing committee notifies, through MSO, the applicant by special notice of the deferral and the grounds for such deferral.
   6. The notice states if the applicant is to provide additional information or a specific release authorization to allow #ORG# representatives to obtain information or documentation,
   7. The notice includes a request for the specific information needed or a signed release/authorization and the deadline for the response, which is not to exceed thirty (30) days.
   8. Failure to respond in a satisfactory manner by the deadline is deemed a withdrawal of the application, unless the credentialing committee determines the failure to respond was caused by circumstances beyond the applicant's control.
   9. The CEO sends the applicant special notice of any deemed withdrawal.
   10. An applicant whose application is withdrawn due to failure to comply with this provision is not entitled to the procedural due process rights provided in the Fair Hearing Plan and the withdrawal is not in itself considered an adverse action reportable under the Health Care Quality Improvement Act and/or relevant state law.
   11. If the credentialing committee conclusions are contrary to those contained in the Department Chair's report, the credentialing committee and Department Chair meets to discuss the differences.
   12. Following these discussions, the credentialing committee or Department Chair may determine to affirm or modify his/her original report. A written summary of the discussions and conclusions is prepared as an addendum to the credentialing committee's report. The credentialing committee/Department Chair discussions provided herein and transmittal of the credentialing committee's report with all supporting documentation is presented to the MEC.
G. Action by the Medical Staff Executive Committee:
   1. The MEC, at its next regular meeting, reviews the application information.
   2. Report of the credentialing committee Chair and other relevant information is made available to the committee.
   3. The MEC may, at its discretion, conduct an interview with the applicant or designate one or more of its members to do so.
   4. A summary of such interview is documented in the minutes of the next MEC meeting subsequent to the interview.

5. The MEC defers action on the application or prepares a written report with recommendations as required above.

H. Effect of MEC action:

1. **Deferral**: Action by the MEC to defer action on the application is followed up within forty-five (45) days with its report and recommendations.

   a. The President of the Medical Staff promptly sends the applicant a special notice through MSO, of the MEC's determination to defer action on the application.

   b. The special notice includes a request for further information if requested by the MEC and/or a signed release/authorization from the applicant and the deadline for the response, which is not to exceed fifteen (15) days.

   c. Failure to respond in a satisfactory manner by the deadline is deemed a withdrawal of the application, unless the MEC determines that the failure to respond was caused by circumstances beyond the applicant's control.

   d. The CEO sends the applicant special notice of any deemed withdrawal.

   e. An applicant whose application is withdrawn due to failure to comply with this provision is not entitled to the procedural due process rights provided in the Fair Hearing Plan and the withdrawal is not in itself considered an adverse action reportable under the Health Care Quality Improvement Act and/or relevant state law.

2. **Favorable Recommended Action**: A MEC recommended action that is favorable to the applicant in all respects is forwarded to the Board Quality Committee (Board) together with supporting documentation.

3. **Adverse Recommended Action**: In the event of an adverse recommended action by the MEC;

   a. The CEO promptly informs the applicant by special notice as per the Fair Hearing Plan and the applicant is then entitled, upon proper and timely request, to the procedural rights provided in said Plan.

   b. For purposes of this section an "adverse recommended action" by the MEC is as defined in the Fair Hearing Plan.

I. Board Action:

   The Board considers the applications recommended by the MEC.

J. Content of Report and Basis for Evaluations, Recommendations and Actions:

1. Each individual and/or committee providing a recommendation or acting on an application has available the full resources of the Medical Staff and #ORG# as well as the authority to use outside consultants as deemed necessary.

2. The report of each individual or group include conclusions as to approval or denial of the application, any applicable limitations regarding appointment, prerogatives, type of clinical privileges, and, if applicable, a recommendation/request that the applicant obtain a physical and/or psychiatric evaluation. If any such conclusions are not included, the reason therefore is stated.

3. All documentation and information received by an individual or group during or as part of the evaluation process is included with the application as part of the individual's central credentials file and as appropriate or requested, transmitted with reports.

4. The reasons for each conclusion, recommendation or action to deny restrict or otherwise limit and for any recommendation/request that the applicant obtains a physical and/or psychiatric evaluation is stated in the report.

K. Notice of Final Decision:

1. The CEO approves and sends special notice of the final Board decision to the applicant via MSO.

2. A decision and notice to appoint includes:

   a. The Staff category to which the applicant is appointed;

   b. The Department and, as applicable, other clinical unit to which the CEO assigned;

   c. The clinical privileges the CEO may exercise;

   d. Any special conditions attached to the appointment; and;

   e. The effective date of appointment.

L. Time Periods for Processing:

1. All individuals and groups required to review or act on an application for Staff appointment and/or clinical privileges should do so in a timely and good faith manner and, except for obtaining required

additional information or for good cause, each application should be processed within the following time periods:

| Individual/Group | Time |
| --- | --- |
| Medical Staff Office | As soon as practical after receipt of the application. |
| Department Chair | As soon as practical after receiving notice file is ready for review. |
| Credentialing Committee | Next scheduled credentialing committee meeting |
| Medical Staff Executive | Next scheduled meeting following the credentialing committee meeting |
| Board of Managers | Next scheduled BOM meeting following recommendation by the MEC |

2. The above schedule is not to be deemed directives such as to create any rights for a practitioner to have an application processed within these precise periods. If the provisions of the Fair Hearing Plan are activated, the time requirements provided there govern the continued processing of the application. If action does not occur at a particular step in the process and the delay is without good cause, the next higher authority immediately proceeds to consider the application and all the supporting information or may be directed by the President of the Medical Staff on behalf of the MEC or by the CEO on behalf of the Board to so proceed.

M.  Conflict Resolution:
1. Whenever the Board or Board determines that it is decided a credentials matter contrary to the recommendation of the MEC, the matter is submitted to a joint advisory council.
    a. The council is composed of an equal number of representatives each from the Medical Staff and the Board appointed respectively by the President of the Medical Staff and the Chair of the Board, for review and report before the final decision is made by the Board or Board, as applicable.
    b. This joint advisory council convenes to review the matter and submit its report to the Board or Board, as applicable, within 60 days after a matter is referred.

N.  Term Of Appointment:
Appointments to the Staff and grants of clinical privileges are for a period of two (2) years. The exceptions to this two (2) year period are as follows:
1. Appointment to the Honorary Staff is for life unless the requirements or obligations set forth for appointment in said category are not satisfied;
2. Each member, except those appointed to the Honorary Staff is requested and required to provide evidence prior to the expiration date, of renewed licensure, DEA And DPS registrations if applicable to his/her practice and professional liability insurance coverage;
3. In order to achieve a system of staggered reappraisal, the appointment of some staff members may be less than two full years;
4. New appointees to the staff are subject to an initial provisional period as provided in this Manual; and
5. Granting of increased privileges to a practitioner may be subject to a provisional period and to review at the end of that period as well as being subject to review at the time of the practitioner's review for reappointment; and
6. The Board, BOM, after considering the recommendations of the applicable Department(s), credentialing committee and/or the MEC may set a more frequent reappraisal period; and
7. Disciplinary action involving appointment and/or clinical privileges may be initiated and taken in the interim between reappraisals under the appropriate provisions of this Procedure and/or the Medical Staff Bylaws; and
8. Case of a practitioner providing professional services by contract employment termination or expiration of the contract/employment may result in a shorter period of appointment or privileges.

O.  Procedures For Reappointment And Reappraisal Of Privileges:
Information Collection And Verification From Practitioner;
1. On or before four (4) months prior to the date of expiration of a practitioner's appointment and/or clinical privileges, MSO staff notifies him/her of the date of expiration and send him/her an application for reappointment/reappraisal.
2. At least three (3) months prior to the expiration date, the practitioner sends the completed application to MSO to include:

  a. Documentation of continuing medical education during the preceding period as required by the State;

  b. Specific request for the clinical privileges requested for the upcoming term, including any basis for changes from the privileges currently held;

  c. Requests for change in Department or other clinical unit or Staff category assignments, if applicable;

  d. The practitioner signs the application and in so doing accepts the same conditions;

  e. If the application and supporting documentation is not received at least three (3) months prior to the expiration date, MSO staff promptly send a special notice requesting receipt within fourteen (14) days from the date of the notice.

3. Other requests may be made of a practitioner, if the evidence at #ORG# of the practitioner's current clinical competence to exercise the privileges requested is not sufficient to permit the applicable Staff and Board authorities to make an informed judgment as to his/her ability to exercise the clinical privileges requested, the practitioner has the burden of providing evidence of clinical competence at other institutions where he holds privileges in such form as may be required by said authorities.

4. Failure, without good cause, to provide the fully completed reappointment/reappraisal application with all of the above information prior to the expiration of the fourteen (14) day grace period results in voluntary resignation of appointment and/or privileges at the expiration of the current term.

  a. The CEO sends the applicant a special notice through MSO, of any deemed withdrawal. An applicant who has their application withdrawn due to failure to comply with this provision is not entitled to the procedural due process rights provided in the Fair Hearing Plan and the withdrawal is not in itself considered an adverse action reportable under the Health Care Quality Improvement Act and/or relevant state law.

5. An applicant who has voluntarily resigned due to failure to comply with this provision is not entitled to the procedural due process rights provided in the Fair Hearing Plan and the resignation is not in itself considered an adverse action reportable under the Health Care Quality Improvement Act and/or relevant state law.

6. If the voluntary resignation occurred during the process of an investigation of the staff member's professional competence or conduct by the credentialing committee, MEC, Board, BOM or other peer review or medical committee, however, such resignation is reported to the National Practitioner's Data Bank, the Board of Medical Examiners and/or any other regulatory agencies as required by law.

P. From Internal Sources:

At the time of reappointment/reappraisal, MSO consolidate for review all available and relevant information regarding the individual's professional and collegial activities, performance and conduct at #ORG# and/or other institutions. Such information forms the basis for recommendations and action, include, without limitation:

1. Professional abilities, clinical judgment, patterns of care and utilization as demonstrated in the findings of quality/risk/utilization assessment, review and improvement activities;

2. Participation in relevant continuing education activities;

3. Number of patient encounters;

4. Sanctions imposed and/or pending and other problems;

5. Health status;

6. Attendance at required meetings;

7. Participation as a Staff official, committee member, chair and proctor;

8. Timely and accurate completion and preparation of medical records;

9. Cooperativeness in working with other practitioners and #ORG# personnel;

10. Compliance with all applicable bylaws, policies, rules and regulations of #ORG# and Medical Staff; and

11. Other pertinent information that may be relevant to the practitioner's activities at other hospitals/healthcare facilities and his/her medical practice outside #ORG#.

12. MSO notifies the Chair of each Department where the practitioner exercised privileges during the last period of appointment and the Chair of each Department where the practitioner is requesting appointment or privileges as to when the reappointment/reappraisal application, the supporting

information and the practitioner's credentials file, or relevant portions thereof, with the information required are available for review.

Q.  Department Evaluation:

1.  The chair of each department the practitioner requests or has exercised privileges review the reappointment/reappraisal application and its supporting information:

    a.  Physician performance and professionalism feedback report.
    b.  Information gathered from internal sources;
    c.  Other pertinent aspects of the practitioner's file and evaluate the information for continuing satisfaction of the qualifications for appointment, the clinical unit(s) and category of assignment and the privileges requested. In the case of a Department Chairman's reappointment, the review is conducted by the medical director.

2.  Department Chair or their designees, may, at their discretion, interview a member applying for reappointment.

3.  The Department Chair notifies the practitioner in writing through MSO if additional information is required.

4.  A special notice is sent to the practitioner requesting the specific information requested by the chair and the response to the request does not exceed fourteen (14) days.

5.  Failure to respond in a satisfactory manner by the deadline is deemed a voluntary resignation of Staff appointment and all clinical privileges; unless the credentialing committee determines that the failure to respond was caused by circumstances beyond the practitioner's control.

6.  Notice of voluntary resignation is forwarded to the credentialing committee, MEC and Board.

7.  The CEO sends the practitioner by special notice of any deemed resignation, and any such resignation carries the same obligation as provided above.

8.  An applicant who has voluntarily resigned due to failure to comply with this provision is not entitled to the procedural due process rights provided in the Fair Hearing Plan and the resignation is not in itself considered an adverse action reportable under the Health Care Quality Improvement Act and/or relevant state law.

9.  Provided the voluntary resignation occurred during the process of an investigation of the staff member's professional competence or conduct by the credentialing committee, MEC, Board of Managers or other peer review or medical committee, such resignation may be reported to the National Practitioner's Data Bank, the Board of Medical Examiners and/or any other regulatory agencies as required by law.

10. Each applicable Department Chair forwards to the credentialing committee a written report, including conclusions regarding reappointment or denial of reappointment, and if applicable, Staff category, department or other clinical unit assignment and clinical privileges, or if no such conclusions are made, the reasons why the decision was not documented.

11. Unapproved reappointments require written communication from the Department Chair to the credentialing committee.

R.  Credentialing Committee Evaluation:

1.  The credentialing committee members review the reappointment/reappraisal application information on the reappointment grid.

2.  Department Chair reports per request,

3.  Qualifications for Staff appointment,

4.  Category of assignment and the privileges requested.

5.  If credentialing committee requires further information, credentialing committee notifies, through MSO, the practitioner by special notice to interview with the credentialing committee or request additional information required and the deadline for the response which is not to exceed fifteen (15) days.

6.  Failure to respond in a satisfactory manner by the deadline is deemed a voluntary resignation of Staff appointment and all clinical privileges; unless the credentialing committee determines that the failure was caused by circumstances beyond the practitioner's control.

7.  The CEO sends the practitioner a special notice of any deemed resignation, and any such resignation carries the same obligations as provided above.

8. An applicant who has voluntarily resigned due to failure to comply with this provision is not entitled to the procedural due process rights provided in the Fair Hearing Plan and the resignation is not in itself considered an adverse action reportable under the Health Care Quality Improvement Act and/or relevant state law.

9. Provided the voluntary resignation occurred during the process of an investigation of the staff member's professional competence or conduct by the credentialing committee, MEC, BOM or other peer review or medical committee, however, such resignation may be reported to the National Practitioner's Data Bank, the Board of Medical Examiners and/or any other regulatory agencies as required by law.

10. The credentialing committee prepares a written report, including conclusions regarding reappointment or denial of reappointment, and, if applicable, staff category, department or other clinical unit assignment, and clinical privileges, or if no such conclusions are made, the reasons therefore.

11. If the credentialing committee's conclusions are contrary to those of a Department Chair, the credentialing committee and the Department Chair engage in the same type of joint discussions as provided. The credentialing committee's report is transmitted with the Department Chair's reports and supporting documentation, as required, to the MEC.

S. Medical Staff Executive Committee:

1. The MEC review the reappointment/reappraisal information, gathered, other pertinent aspects of the practitioner's file as needed, the chair and credentialing committee's reports and all other relevant information available.

2. The MEC may at its discretion, conduct an interview with the applicant or designate one of more of its members to do so.

3. A summary of such interview is documented in the minutes of the next MEC meeting subsequent to the interview.

4. The MEC may defer action or prepare a written report with recommendations as required above.

5. May refer the application for reappointment to credentialing committee for re-evaluation if additional information is requested.

## Action by the Medical Executive Committee:

1. Deferral: Action by the MEC to defer action n the application is followed up within fortyfive (45) days with its report and recommendations.

2. The President of the Medical Staff promptly sends the applicant a special notice through MSO, of the MEC determination to defer, action on the application.

3. The special notice includes a request for further information if requested by the MEC and/or a signed release/authorization from the applicant and the deadline is not to exceed fifteen (15) days.

4. Failure to respond in a satisfactory manner by the deadline is deemed a withdrawal of the application, unless the MEC determined the failure to respond was caused by circumstances beyond the applicant's control.

5. The CEO sends the applicant special notice of any deemed withdrawal.

6. An applicant whose application withdrawn due to failure to comply with this provision is not entitled to the procedural due process rights provided in the Fair Hearing Plan and the withdrawal is not in itself considered an adverse action reportable under the Health Care Quality Improvement Act and/or relevant state law.

Final Processing:

Final processing of all reappointment/reappraisals applications is to be carried out in accordance with the procedure set above. For purposes of reappointment/reappraisal applications, the terms "applicant" and "application" as used in said Sections mean, respectively, "practitioner" and "reappointment/reappraisal."

Bases for Conclusions, Recommendations And Action:

Each individual and committee reviewing or acting on a reappointment/reappraisal application has available the full resources of the Medical Staff and #ORG# as well as the authority to use outside consultants as deemed necessary. The report of each individual or committee required to act on a reappointment/reappraisal application state the reasons for each adverse conclusion or recommendation made or action taken, and said reasons is recorded in the Medical Staff minutes.

If the delay is attributable to the practitioner's failure to provide the information required above or other information required by a reviewing individual or committee, his Staff appointment and/or clinical privileges terminates on the expiration date as provided above unless expressly extended as provided therein.

## SYSTEMS AND PROCEDURES FOR DELINEATING CLINICAL PRIVILEGES:

*Exercise of Privileges in General:*

1. A practitioner providing clinical services at #ORG# either by virtue of Staff appointment, as described above or as a temporary privilege situation may, in connection with such practice and except as otherwise provided in above in an emergency, exercise only those clinical privileges specifically granted to him pursuant to this procedure.
2. There may be attached to any grant of privileges to individual practitioner special requirements for consultation as a condition to the exercise of particular privileges.
3. Each practitioner provides or arranges for continuous medical care for his/her patients at #ORG#.
4. Any physician who is covering for a practitioner must be a member of the Medical Staff.
5. When dealing with a problem or condition outside a practitioner's training and usual area of practice, or when required by the rules or other policies of the Staff, any of its clinical units, or #ORG#, each practitioner obtains appropriate consultation/proctoring or refers the case to another qualified practitioner when necessary for the safety of his patient.
6. A practitioner may be granted clinical privileges in one or more of the departments of the Staff, and his/her exercise of privileges within the jurisdiction of any department is always subject to the rules and regulations of that unit and the authority of the Department Chairman.

*Restrictions on Privileges Temporarily Waived in an Emergency:*

For existing medical staff members, in an emergency, any medical staff member with clinical privileges is "temporary privileged" to provide any type of patient care necessary as a life-saving measure or to prevent serious harm, regardless of his or her current clinical privileges, if the care provided is within the scope of the individual's licenses. In facilities with approved graduate medical education programs, properly supervised members of the house staff may provide such emergency care. Described above and #ORG# Policy.

*Basis for Privilege Determinations:*

1. Clinical practice privileges are granted in accordance with prior and continuing education and training, prior and current experience, utilization practice patterns, current health status, and demonstrated current competence and judgment as documented and verified in each practitioner's credentials file.
2. Additional factors that may be used in determining privileges are patient care needs for and facility capability to support the type of privileges being requested by the applicant and the availability of qualified coverage in his absence.
3. The basis for privilege determination for practitioners in connection with reappraisal, including conclusion of the provisional period, or with a requested change in privileges also include observes clinical performance, documented results of the Staff's quality/risk/utilization assessment, review and improvement activities.
4. In the case of additional privileges requested, verified evidence of appropriate training and experience supportive of the request is provided.

*Special Conditions for Practitioners Not Appointed to the Staff:*

A practitioner who is not a member of the Staff may be granted clinical privileges pursuant to the procedures set above. Any application for and grant of privileges pursuant to this section is subject to the same qualifications, terms and conditions as are set forth in this Manual for Medical Staff members or applicants.

*Principles Governing Allied Health Professionals:*

The principles governing the granting to and performance by Allied Health Professionals (AHPs) of specified patient care services are set forth in the Allied Health Professional guidelines.

*Responsibility to Define Approach to Delineating Privileges:*

1. Each department and section, as applicable, defines, in writing the clinical privileges required in their department.
2. The definitions and delineating instruments periodically reviewed and revised as necessary to reflect new procedures, instrumentation, treatment modalities and like advances or changes.
3. When definitions/delineation instruments are revised, by additions or deletions or the adoption of new forms, all practitioners holding privileges in that Department/Section, as appropriate to the circumstances, complete the new forms, request and are processed for privileges added, or comply with the fact that a privilege was deleted.

*Procedure for Delineating Privileges Requests:*

1. Each application for appointment and reappointment/reappraisal contain a request for the specific clinical privileges desired by the practitioner.
2. Specific requests are also submitted for privileges to be exercised as described above and for modifications of privileges in the interim between reappointment/reappraisals.

*Processing Requests:*

1. A request for clinical privileges, except one for temporary privileges, is processed according to the procedures outlined if it is by a practitioner not currently a member of the Staff or one not having current privileges or according to those outlined if it is by a current Staff member or a practitioner with a current grant of privileges. Requests for temporary privileges are processed.

*Temporary Privileges:*

*Circumstances;*

*The CEO, or his designee, may grant temporary privileges under the following circumstances:*

1. Pending Application:
   a. To an applicant for initial appointment when the information available supports a favorable recommendation regarding the practitioner's application for appointment and privileges, but only after receipt of a completed application for Staff appointment including a request for specific privileges, and completion of the application verification process including receipt of all possible responses.
   b. Temporary privileges may be granted in this circumstance for an initial period of one hundred twenty (120) days. Under no circumstances does the initial one hundred twenty day (120) time period be extended. Under no circumstances may temporary privileges be granted if the applicant has not responded in a satisfactory manner to a request for clarification of a matter or for additional information or if the recommended action of the Department Chair, credentialing committee, MEC or Board is adverse in any respect.
   c. Care of Specific Patient: To a practitioner for the care of a specific patient but only after receipt of an application, which includes a request for the specific privileges desired; written concurrence of the applicable Department Chair,, President of the Medical Staff, medical director, and the CEO; at least telephonic verification of appropriate licensure and receipt of a copy of current DEA and state controlled substance registrations and professional liability insurance coverage; a fully positive reference specific to the privileges being requested from a responsible medical staff authority, whose professional skills and competence are known directly or by reputation to the President of the Medical Staff or to some other Medical Staff authority.
   d. Temporary privileges of this nature may not be granted in more than one (1) instance in any twelve (12) month period after which the practitioner applies for staff appointment. When there is not a #ORG# Staff member in the particular specialty requested, there may be up to three (3) requests per year, after which the practitioner apply for Staff appointment.
2. Temporary Membership: Temporary Membership may be granted on a case by case basis due to patient care needs only after receipt of an application, including a request for specific privileges, completion of the application verification process, including receipt of all possible responses, and a fully affirmative recommendation by each applicable Department Chair, President of the Medical Staff and the CEO. Temporary Membership may not exceed one hundred and twenty (120) consecutive days.
3. Consultative Privileges: Consultative privileges include the authority to provide consultation and/or education regarding patient management at the request of a member of the Active or Courtesy Staff only after: receipt of an application, which includes a request for the specific privileges desired; written concurrence of

the applicable Department Chair, medical director, and CEO; at least telephonic verification of appropriate licensure and receipt of a copy of current DEA and, state controlled substances, registrations and professional liability insurance coverage; a fully positive reference specific to the privileges being requested from a responsible medical staff authority at the practitioner's current principal hospital, preferably whose professional skills and competence are known directly or by reputation to the medical director, or to some other Medical Staff authority.

4. Procedural Privileges: Procedural privileges include:
   a. Privileges for a specific procedure or privileges to assist and/or provide guidance in the performance of a specific medical or surgical diagnostic or therapeutic procedure at the request of a Medical Staff member;
   b. As part of an organized course sponsored by #ORG# only after receipt of an application, which includes a request for the specific privileges desired; written concurrence from the applicable Department Chair, President of the Medical Staff, medical director, and CEO;
   c. At least telephonic verification of appropriate licensure and receipt of a copy of current DEA and state controlled substances, registrations and professional liability insurance coverage;
   d. A fully positive reference specific to the privileges being requested from a responsible medical staff authority at the practitioner's current principal hospital, preferably whose professional skills and competence are known directly or by reputation to the President of the Medical Staff, or to some other Medical Staff authority.

5. Termination:
   a. The CEO, upon recommendation of the Department Chair, President of the Medical Staff, medical director and/or BOM may terminate any or all of a practitioner's temporary privileges. Temporary privileges granted are automatically terminated in the event of an adverse recommendation or action by the MEC or the BOM. The CEO sends the practitioner special notice of any deemed termination.
   b. In the event of termination of temporary privileges, the practitioner's patients are assigned to another practitioner by the applicable Department Chair.

*Rights of the Practitioner:*

A practitioner is not entitled to the procedural due process rights afforded by the Fair Hearing Plan because his/her request for temporary privileges is refused in whole or in part or because all or any portion of his temporary privileges are terminated, not renewed, restricted, suspended, or otherwise limited in any way, unless such action is based on the professional competence or professional conduct of the practitioner which affects, or could affect, adversely the health or welfare of a patient.

*Emergency Privileges:*

*Circumstances:*

1. #ORG# may grant emergency privileges to a Licensed Independent Practitioner (LIP) when the Emergency Management Plan has been activated and #ORG# is unable to handle the immediate patient needs, as follows: Emergency privileges on a case by case basis may be approved when the Emergency Management Plan has been activated and the need mandates immediate authorization to practice to fulfill an important patient care need, until the Emergency Management Plan has been deactivated.
2. Any medical staff member is permitted to provide any type of patient care necessary as a life-saving measure or to prevent serious harm, regardless of his or her medical staff status or clinical privileges; provided the care is within the scope of the individual's license.
3. The CEO, COO, medical director, President of the Medical Staff or the Chair of Emergency Services may grant emergency privileges upon the presentation of any of the following:
   a. Current picture facility identification (ID) badge; or
   b. Current license to practice and a valid picture ID issued by a state, federal or regulatory agency; or
   c. Identification indicating the individual is a member of a Disaster Medical Assistance Team (DMAT); or
   d. Identification that the individual has been granted authority to render patient care in emergency circumstances, such authority having been granted by a federal, state or a municipal entity.
   e. Photocopies of the above listed documents should be made and retained.
   f. MSO staff is available to review and verify the above information according to the initial application process as soon as possible and preferably prior to the LIP being granted temporary privileges.

## CONCLUSION AND EXTENSION OF PROVISIONAL PERIOD

*Applicability and Duration:*

1. New appointments to the Staff and grants of initial clinical privileges are subject to a provisional period of twelve (12) months, unless extended pursuant.
2. Grants of increased privileges to an existing Staff member or practitioner with privileges are also subject to a provisional period as defined by the Department Chair with concurrence by the credentialing committee, subject to the approval of the MEC and the BOM.
3. During this period, a practitioner's performance is reviewed and evaluated in accordance with the Initial Competency Evaluation Policy/Provisional of #ORG#.

*Status and Privileges during Provisional Period:*

1. During the first year of Initial Appointment (provisional period), a practitioner meets all requirements, may exercise all of the prerogatives, and fulfills all obligations of their Staff category. He/she may exercise all clinical privileges granted to him/her under such conditions of observation or supervision as are established.
2. A practitioner's exercise of prerogatives and clinical privileges during the provisional period is subject to Initial Competency Evaluation Policy/Provisional of #ORG#.

*Role of Proctor:*

The role of the proctor is to observe, review and report to the Department Chair on the clinical performance and competence of the practitioner. The proctor is not deemed to have entered into a physician patient relationship on any case on which he is acting as a proctor, not speak to or examine the patient, and does not consult with, supervise, assist, advise, scrub in with, or otherwise intervene in the care being provided by the practitioner being proctored.

If the proctor has concerns regarding the care he is observing being provided by the proctored practitioner, he so informs the practitioner's Department Chair as soon as practical.

*Action Required:*

1. At least ninety (90) days prior to the end of the provisional period for a practitioner with increased privileges, MSO completes the necessary documentation for the chair of the department to begin the review process.
2. For purposes of concluding the provisional period, the term "reappointment/reappraisal" as used in said Sections mean "conclusion of the provisional period."

## REVIEW AND AMENDMENT

This Credentialing Procedures are reviewed at two (2) year intervals by the credentialing committee, and may be reviewed more frequently when deemed necessary by the appropriate Medical Staff or Board authorities. Suggestions for changes in the procedure referred to the credentialing committee, which present its recommendations in a timely fashion to the MEC.

The procedural may be amended, in whole or in part, or a new one proposed by the affirmative vote of a majority of the MEC members present at a regular or special meeting at which a quorum is present and the affirmative vote of a majority of the Board members present at a regular or special meeting at which a quorum is present. Amendments to or a new procedure becomes effective upon the affirmative vote of the BOM.

**REFERENCES:**

**FORMS:**

**EQUIPMENT:**

**APPROVALS:**

| NAME | TITLE | DATE |
|---|---|---|
| #APPROVER# | #APPRTITLE# | #APPRDATE# |

**Policy and Procedures**

Policy#: #POLNUM#
Location: #LOC#
Originating Department: #ORIGDEPT#
Effective Date: #EFFDATE#
Expiration Date: #EXPDATE#

## TITLE: MEDICAL STAFF PEER REVIEW

### POLICY STATEMENT:

It is the policy of #ORG# to establish that #ORG#, through the activities of its medical staff, assesses the Ongoing Professional Practice Evaluation of individuals granted clinical privileges and uses the results of such assessments to improve care and, when necessary, performs an evaluation.

### INTENT AND SCOPE:

It is the intent of this policy to:

1. Monitor and evaluate the ongoing professional practice of individual practitioners with clinical privileges;
2. Create a culture with a positive approach to peer review by recognizing physician excellence as well as identifying improvement opportunities;
3. Perform focused professional practice evaluation when potential physician improvement opportunities are identified;
4. Promote efficient use of physician and quality staff resources;
5. Provide accurate and timely performance data for physician feedback, Ongoing and Focused Professional Practice Evaluation and reappointment; and,
6. Assure that the process for peer review is clearly defined, fair, defensible, timely and useful.

This policy applies to any practitioner credentialed under medical staff bylaws.

### DEFINITIONS:

I. Peer review—is the evaluation of an individual practitioner's professional performance and includes the identification of opportunities to improve care. Peer review differs from other quality improvement processes in that it evaluates the strengths and weaknesses of an individual practitioner's performance, rather than appraising the quality of care rendered by a group of professionals or a system.

Peer review is conducted using multiple sources of information including:
1) the review of individual cases, 2) the review of aggregate data for compliance with general rules of the medical staff and, 3) clinical standards and use of rates in comparison with established benchmarks or norms.

The individual's evaluation is based on generally recognized standards of care. Through this process, practitioners receive feedback for personal improvement or confirmation of personal achievement related to the effectiveness of their professional practice. ACGME is used as a reference for this process.

II. Peer—is an individual practicing in the same profession and who has expertise in the appropriate subject matter. The level of subject matter expertise required to provide meaningful evaluation of a practitioner's performance determines what "practicing in the same profession" means on a case-by-case basis. For example, for quality issues related to general medical care, a physician (MD or DO) may review the care of another physician. For specialty-specific clinical issues, such as evaluating the technique of a specialized surgical procedure, a peer is an individual who is well-trained and competent in that surgical specialty.

III. Peer Review Body—is designated to perform the initial review by the Medical Executive Committee or its designee determines the degree of subject matter expertise required for a provider to be considered a peer for all peer reviews performed by or on behalf of the facility. The initial peer review body is the Medical Executive Committee

IV. Ongoing Professional Practice Evaluation (OPPE)—the routine monitoring and evaluation of current competency for current medical staff. These activities comprise the majority of the functions of the ongoing peer review process and the use of data for reappointment.

V. Professional Practice Evaluation (PPE)—the establishment of current competency for new medical staff members, new privileges and or concerns from OPPE. These activities comprise what is typically called proctoring or focused review depending on the nature of the circumstances.

## General Information:

I. All peer review information is privileged and confidential in accordance with medical staff and facility bylaws, state and federal laws, and regulations pertaining to confidentiality and non-discoverability.

II. The involved practitioner receives provider-specific feedback on a routine basis.

III. The medical staff uses provider-specific peer review results in making its recommendations to the facility regarding the credentialing and privileging process and, as appropriate, in its performance improvement activities.

IV. #ORG# keeps provider-specific peer review and other quality information concerning a practitioner in a secure, locked file. Provider-specific peer review information consists of information related to:
A. Performance data for all dimensions of performance measured for that individual physician,
B. The individual physician's role in sentinel events, significant incidents or near misses,
C. Correspondence to the physician regarding commendations, comments regarding practice performance, or corrective action.

V. Only the final determinations of the MEC and any subsequent actions are considered part of an individual provider's quality file.

VI. Peer review information in the individual provider quality file is available only to authorized individuals who have a legitimate need to know this information based upon their responsibilities as a medical staff leader or facility employee. However, they shall have access to the information only to the extent necessary to carry out their assigned responsibilities. The Medical Director determines that only authorized individuals have access to individual provider quality files and that the files are reviewed under the supervision of the Medical Director. Only the following individuals have access to provider-specific peer review information and only for purposes of quality improvement:
A. The specific provider;
B. The Medical Director for purposes of considering corrective action;
C. Medical staff department chairs (for members of their departments only) to conduct OPPE;
D. Members of the medical executive committee, credentials committee and medical staff services professionals for purposes of considering reappointment or correction action.
E. Quality and support staff supporting the peer review process;
F. Individuals surveying for accrediting bodies with appropriate jurisdiction, e.g. state/federal regulatory bodies; and
G. Individuals with a legitimate purpose for access as determined by the Board of Managers.

H.  The administrator when involvement in the process of immediate formal corrective action for purposes of summary suspension as defined by the medical staff bylaws.

VII.  No copies of peer review documents are created and distributed unless authorized by medical staff policy or bylaws, the MEC, the Board for purposes of deliberations regarding corrective action on specific cases.

VIII.  Circumstances requiring peer review:

Peer review is conducted on an ongoing basis and reported to the appropriate committee for review and action.

IX.  Circumstances requiring external peer review:

Either the Department Chair, or the MEC make determinations on the need for external peer review to be funded by the facility. No practitioner can require the facility to obtain external peer review if it is not deemed appropriate by the MEC or the Department Chair. Circumstances requiring external peer review may include but are not necessarily limited to:
A.  Litigation—when dealing with the potential for a lawsuit.
B.  Validation of internal peer review findings.
C.  Ambiguity—when dealing with vague or conflicting recommendations from internal reviewers or medical staff committees and conclusions from this review directly impacts a practitioner's membership or privileges.
D.  Lack of internal expertise—when no one on the medical staff has adequate expertise in the specialty under review; or when the only practitioners on the medical staff with that expertise are determined to have a conflict of interest regarding the practitioner under review as describe above. External peer review takes place if this potential for conflict of interest cannot be appropriately resolved by the medical executive committee or governing board.
E.  Miscellaneous issues—when the medical staff needs an expert witness for a fair hearing, for evaluation of a credential file, or for assistance in developing a benchmark for quality monitoring. In addition, the medical executive committee or governing board may require external peer review in any circumstances deemed appropriate by either of these bodies.

X.  Participants in the review process:

Participants in the review process are selected according to the medical staff policies and procedures. The work of all practitioners granted privileges are reviewed through the peer review process. Clinical support staff participates in the review process if deemed appropriate. Additional support staff participates if such participation is included in their job responsibilities. The peer review body considers the views of the person whose care is under review prior to making a final determination regarding the care provided by that individual providing that individual response in the timeframe outlined.

In the event of a conflict of interest or circumstances that would suggest a biased review beyond that described above, the MEC replace, appoint or determine who participates in the process so that bias does not interfere in the decision-making process.

XI.  Selection of Physician Performance Measures

Measures of physician performance are selected to reflect the evaluation process.

XII.  Thresholds for Professional Practice Evaluation:

If the results of Ongoing Professional Practice Evaluation indicate a potential issue with physician performance, the MEC may initiate a focused evaluation to determine if there is problem with current competency of the physician for either specific privileges or for more global dimensions of performance. These potential issues may be the result of individual case review or rule or rate indicators. The thresholds for Professional Practice Evaluation are described in the acceptable targets for the medical staff.

XIII. Individual Case Review

Peer review is conducted by the medical staff in a timely manner. The goal is for routine cases to be completed within ninety (90) days from the date the chart is reviewed by the quality/risk staff and complex cases to be completed within one hundred and twenty (120) days. Exceptions may occur based on case complexity or reviewer availability. The timelines for this process must be outlined. The rating system for determining results of individual case reviews is defined and available in the review forms.

XIV. Rate and Rule Indicator Data Evaluation:

The evaluation of aggregate physician performance measures via either rate or rule indicators are conducted on an ongoing basis by the MEC or its designee.

XV. Oversight and Reporting:

Direct oversight of the peer review process is accomplished by the MEC. The MEC reports to the Board of Managers at least quarterly.

XVI. Statutory Authority:

This policy is based on the statutory authority of the Health Care Quality Improvement Act, 42 U.S.C. Section 11101, et seq. All minutes, reports, recommendations, communications and actions made or taken pursuant to this policy are deemed to be covered by such provisions of federal and state law providing protection to peer review related activities. Documents, including minutes and case review materials, prepared in connection with this policy are labeled with language consistent with the following:

XVII."Statement of confidentiality"

Medical Committee/Medical Peer Review Committee Document—Confidential and Privileged pursuant to the #STATE# Occupations Code (Medical Peer Review Committee Privilege), of the #STATE# Health and Safety Code (Medical Committee Privilege), the Health Care Quality Improvement Act, 42 U.S.C. Section 11101, et seq. Do Not Copy or Distribute—Use Exclusively for Committee Purposes.

**PROCEDURES:**

I. Participation Fairness Doctrine:
A member of the medical staff requested to perform peer review may have a conflict of interest if they may not be able to render an unbiased opinion. An absolute conflict of interest would result if the physician is the provider under review. Relative conflicts of interest are either due to a provider's involvement in the patient's care not related to the issues under review or because of a relationship with the physician involved as a direct competitor, partner or key referral source that exist above the normal system physician connections. It is the obligation of the individual reviewer to disclose to the peer review committee any potential conflict(s).
It is the responsibility of the peer review body is to determine on a case by case basis if a relative conflict is substantial enough to prevent the individual from participating. When a potential relative conflict is identified, the MEC chair is informed in advance and makes the determination if a substantial conflict exists. When either an absolute or substantial relative conflict is determined to exist, the individual may not participate or be present during peer review body discussions or decisions other than to provide specific information requested as described in the Peer Review Process.

**REFERENCES:**

**FORMS:**

**EQUIPMENT:**

**APPROVALS:**

| **NAME** | **TITLE** | **DATE** |
|---|---|---|
| #APPROVER# | #APPRTITLE# | #APPRDATE# |

**Policy and Procedures**

Policy#:  #POLNUM#
Location:  #LOC#
Originating Department:  #ORIGDEPT#
Effective Date:  #EFFDATE#
Expiration Date:  #EXPDATE#

## TITLE: MEDICAL STAFF PROCTORING

### POLICY STATEMENT:

It is the policy of the Medical Staff of #ORG# to provide a mechanism of proctoring members of the medical staff.

### INTENT AND SCOPE:

It is the intent of this policy to establish a mechanism of proctoring for Medical Staff members.

### DEFINITIONS:

### GENERAL INFORMATION:

Proctoring may be requested in any of the following situations:

I. Initial applicant to the staff may be proctored prior to having specific privileges granted.

II. Current members of the Staff may be proctored when additional or expanded privileges are requested.

III. Proctoring may be utilized as a mechanism to evaluate privileges that are being contested.

IV. Proctoring may be utilized when there is a question about the practitioner's skill level.

### PROCEDURES:

I. When it has been determined that a proctoring situation exists, the Medical Executive Committee:
   A. Assigns a proctor(s) to the practitioner. The department and the staff members involved in the proctoring must be in agreement on the proctors to be used during the process. The staff member being proctored may suggest a proctor to the committee.
   B. Determines the number of cases to be evaluated. This number must be uniform among applicants for a specific procedure/process.

II. If the proctoring is to occur in a different department, the Medical Executive Committee appoints two proctors, one from the delineating department and one from the practitioner's sponsoring department.

III. The practitioner being proctored is responsible for notifying the proctor prior to the time the proctoring is to occur. If the practitioner fails to notify the proctor, the Medical Executive Committee does not accept the case for review.

IV. The proctor should be immediately available at the time a procedure is being performed by the practitioner.

V. At the time of each case, the proctor completes a case review form and submits it to the Medical Director. If there are two proctors, each signs the form. If there is a disagreement between proctors, two separate forms are completed and submitted to the Medical Director.

VI. The Medical Director compiles the results of the proctoring and submits a report to the Medical Executive Committee.

VII. The Medical Executive Committee reviews the results and makes a recommendation.

**REFERENCES:**

**FORMS:**

**EQUIPMENT:**

**APPROVALS:**

| NAME | TITLE | DATE |
|------|-------|------|
| #APPROVER# | #APPRTITLE# | #APPRDATE# |

**Policy and Procedures**

Policy#:  #POLNUM#
Location:  #LOC#
Originating Department:  #ORIGDEPT#
Effective Date:  #EFFDATE#
Expiration Date:  #EXPDATE#

## TITLE: MEDICATION MANAGEMENT: LABELING REQUIREMENTS

### POLICY STATEMENT:

It is the policy of #ORG# to reduce the risk of medication errors by providing #ORG#-wide standards for safe medication practices, including medication labeling, documentation and related communications.

### INTENT AND SCOPE:

The policy is intended to provide a guideline for the labeling of medications that are prepared outside of the pharmacy and removed from their original package/container and not administered immediately. Medications are labeled consistently and with all applicable laws or regulations as well as accepted standards of practice. These standards apply to all ambulatory settings and to include anesthesia and procedural areas. This policy applies to all employees (contract and non-contract) and health care professionals participating in the medication use process.

### DEFINITIONS:

I.  Container/Storage Device: Any container used to store or administer a medication. Examples of such containers include but are not limited to plastic bags, syringes, bottles, boxes, basins, and medicine cups.

II.  Expiration Date or "Beyond Use Date": The day/time by which the product should be administered or used on a patient.

III.  Label: Printed information providing identification of the contents of the container or storage device. A label can be part of the original package, a separate document affixed to the container/storage device, or a combination of both.

IV.  Immediate Use: The administration of a medication/solution without any delay or moving to another function prior to administration.

V.  Delayed Administration: When any amount of time elapses from the time a medication/solution is prepared in which the preparer moves to another function prior to administration.

VI.  Individualized Dose: When a medication/solution is ordered by a practitioner in a specific dose for a specific patient in whom the dose may vary from patient to patient.

VII.  Qualified Individual: Individual within scope of practice that either administers or verifies medication name/dosage.

VIII.  Medication:
    A.  Any prescription;
    B.  Over-the-counter (OTC) drug;

C. Sample medication;
D. Herbal remedies;
E. Vitamins
F. Nutriceuticals;
G. Vaccines;
H. Diagnostic/contrast agents;
I. Radioactive medications
J. Any parenteral preparation;
K. Blood derivatives;
L. Intravenous solutions (plain, with electrolytes and/or medications);
M. Any product designated by the FDA as a drug.

IX. Solution:
A. Chemicals;
B. Reagents;
C. Saline;
D. Sterile water;
E. Lugol's solution;
F. Dyes (to include radiopaque);
G. Preservatives.

## GENERAL INFORMATION:

I. Standardized, pre-printed labels are to be ordered.

II. Customized labels meet labeling guidelines when utilizing the procedure process that is outlined in the policy.

III. Labeling of items removed from their original container/package occurs at the time the medication is being prepared, even if there is only one medication being prepared. Use of pre-labeled packaging from the manufacturer is unacceptable.

IV. Any unlabeled or partially labeled medication or solution is discarded immediately.

V. When two (2) or more people participate in the preparation and/or the administration of the medication or solution, verification is required by two people. However, if the same person prepares and administers a medication or pours and uses the solution, verification by two people is not required. If one person prepares the medication or solution that is not used immediately, it will need to be labeled but verification is not required.

VI. Upon a shift change or break relief (change in personnel involved with labeled medications, such as in the operating room or procedural areas), medications/solutions and their labels are reviewed by entering and exiting persons.

VII. Attaching the original container (i.e. vial/amp, etc) to the final container is unacceptable.

## PROCEDURES:

I. Labeling occurs as follows:
A. Sterile field in a procedural area:
1. Medications/solutions placed onto a sterile field within a procedural area are labeled with the following information:
a. Drug name;
b. Strength/amount—concentration (if not apparent from container);
c. Expiration date (if not used within 24 hours);
d. Expiration time if less than 24 hours; and,

   e. Date prepared and the diluents for all compounded IV admixtures;

   f. These medications are discarded at the end of the procedure.

   g. Original medication/solution containers remain available until the conclusion of the procedure.

 2. Other areas (not a sterile field in a procedural area)

   a. Medications/solutions removed from their original package/container and transferred to another container is labeled with the following information:

    i. Drug Name;

    ii. Drug strength/amount (if not apparent from container);

    iii. Diluents for compounded IV admixtures;

    iv. Date and time prepared;

    v. Initials of preparer;

    vi. Expiration date/time if expiring within 24 hours;

 3. In addition to the above information, the following information is included if the medication/solution is individualized for a specific patient or if the preparer is not the same person as the one administering:

   a. Patient name;

   b. Patient location;

   c. Directions for use;

   d. Cautionary statements (i.e. requires refrigeration, IM only).

**REFERENCES:**

**FORMS:**

**EQUIPMENT:**

**APPROVALS:**

| NAME | TITLE | DATE |
| --- | --- | --- |
| #APPROVER# | #APPRTITLE# | #APPRDATE# |

**Policy and Procedures**

Policy#:  #POLNUM#
Location:  #LOC#
Originating Department:  #ORIGDEPT#
Effective Date:  #EFFDATE#
Expiration Date:  #EXPDATE#

## TITLE: MEDICATION RECONCILIATION

### POLICY STATEMENT:

It is the policy of #ORG# to establish a process and guidelines for Medication Reconciliation.

### INTENT AND SCOPE:

The policy is intended to provide a guideline to accurately and completely reconcile patient medications for our patients. This policy applies to all employees (contract and non-contract) and health professionals participating in patient care.

### DEFINITIONS:

I. Medication Reconciliation—a process for obtaining and documenting a complete list of the patient's current medications at the point of entry to the facility and with the involvement of the patient (if condition allows). This process includes a comparison of the medications the organization provides to those on the list.

II. Communication upon transfer—a complete list of the patient's medications is communicated to the next provider of service when a patient is referred or transferred to another setting, service, practitioner or level of care within or outside the organization.

III. Discrepancies—e.g. omissions, duplications, potential interactions.

IV. Licensed Independent Practitioner—M.D., D.O., Nurse Practitioner and/or Physician Assistant.

V. Outpatient Procedure: Patients are scheduled for non-invasive and/or invasive procedures that may require moderate sedation and/or local anesthetic that are not considered a physician visit.

VI. Outpatient Clinic: Patients are scheduled with a physician for an examination that does not include a defined coded procedure.

### GENERAL INFORMATION:

I. Medication Reconciliation is required upon admission/entry into an ambulatory setting, upon transfer to another level of care and at discharge.

II. When possible, the patient is involved in obtaining a list of their current home medications.

III. For Entrance into #ORG# (e.g. admission, ASC, clinic, procedure):
   A. The Medication Reconciliation Form is utilized by the following to include, but are not limited to: all ASC admissions and any areas where procedures are performed and medications are relevant to their services.
   B. The Outpatient Medication Reconciliation Form and/or other approved computer generated processes may be utilized for all other clinic areas.

IV. The admitting Nurse that completes the initial patient evaluation; creates a complete list of the patient's current medications on the appropriate Medication Reconciliation form. The Licensed Vocational Nurse may assist in obtaining the list of current home medication.

V. Staff performing admission and/or pre-procedure assessments documents the medication list the patient was on prior to arrival.

VI. The medications that are ordered for the patient while under the care of the organization are compared to those on the medication list and any discrepancies are resolved.

VII. Medical Assistants in the outpatient clinical setting may assist in obtaining the patient's medication list on the Medication Reconciliation form, but the clinic nurse/licensed independent practitioner is responsible for assessing the accuracy of the information.

VIII. Sources for establishing a current list of medications may vary depending upon the specific site and/or circumstances depending upon the patient's condition and the reliability of the patient to provide the information. Sources that may be utilized by providers and staff include, but are not limited to:
  A. Patient
  B. Family
  C. Admission History & Physical
  D. Discharge summaries
  E. Patient Pharmacies
  F. Medication Administration Record (MAR)

## PROCEDURES:

I. Admission Process:
  A. Procedural
    1. STEP 1
      a. Upon arrival, staff obtains, from the patient (whenever possible), a complete list of patient's home medications that include the name, dosage, frequency, route, and last dose. This data is placed on the Medication Reconciliation form. The nurse documents the name and phone number of the person providing the information, if other than the patient.
      b. All reasonable attempts are made to obtain an accurate and complete medication list.
      c. The Registered Nurse documents if the patient denies taking any home medication.
      d. The signed Medication Reconciliation form is placed in the orders section of the patient's chart.
    2. STEP 2
      a. The licensed independent practitioner writing admission orders is responsible for reviewing the home medication list and any discrepancies are resolved (reconciled).
      b. The initial Medication Reconciliation is completed prior to the procedure; unless unable to obtain.
      c. Whenever a new medication is prescribed/ordered for a patient, policy regarding Safe Medication Practices is followed.

II. Transfer/Transition Process:
  A. Upon any transfer/transition of care, the licensed independent practitioner documents new orders. Medications are reconciled against the patient's home medication list and their current medication list.
  B. The next provider of service checks the medication reconciliation list again to make sure it is accurate and in agreement with any new medication to be ordered/prescribed.

III. Discharge Process:
  A. A complete list of medications is provided to the patient upon discharge from #ORG#. The medication list should include all medications that the patient is expected to continue.
  B. The medication list is provided to the patient's physician or long-term care facility, wherever the patient is receiving care, if the provider information is available.

IV. Outpatient Process:
    A.   Upon a visit to the outpatient setting an initial medication list is obtained and reconciled with any medications that may be given during that visit, to include radiology.
    B.   If the patient's home medication list does change, there needs to be an updated medication list provided to the patient and the provider of care as appropriate.
    C.   If the patient's home medication list does not change, there is no "Discharge Reconciliation" requirement.

**REFERENCES:**

**FORMS:**

**EQUIPMENT:**

**APPROVALS:**

| NAME | TITLE | DATE |
| --- | --- | --- |
| #APPROVER# | #APPRTITLE# | #APPRDATE# |

**Policy and Procedures**

Policy#:  #POLNUM#
Location:  #LOC#
Originating Department:  #ORIGDEPT#
Effective Date:  #EFFDATE#
Expiration Date:  #EXPDATE#

## TITLE: MODERATE SEDATION

### POLICY STATEMENT:

It is the policy of #ORG# that moderate sedation is safely utilized, that essential items required in an emergency situation be immediately available and that approved processes for monitoring and discharging patients who receive moderate sedation when undergoing intervention procedures be followed. All moderate sedation performed at #ORG# is done in accordance with the American Society of Anesthesiology guidelines, CMS Conditions for Coverage, state licensing requirements and accreditation standards. This policy does not refer to pain-control, patients that are mechanically ventilated, end of life care, or emergent procedures.

### INTENT AND SCOPE:

This policy is intended to communicate #ORG# approved standards in the use of moderate sedation and applies to all Medical Staff and Registered Nursing Staff (contract and non-contract) in settings where procedural moderate sedation may occur including but not limited to the ASC, Endoscopy, Cardiology, Radiology, and clinics.

### DEFINITIONS:

Definitions of the five (5) levels of sedation and anesthesia include the following:

I. Minimal Sedation (anxiolysis): A drug-induced state during which patient responds normally to verbal commands. Although cognitive function and coordination may be impaired, ventilatory and cardiovascular functions are unaffected.

II. Moderate Sedation/Analgesia: A drug-induced depression of consciousness during which patients respond purposefully (reflex withdrawal from a painful stimulus is not considered a purposeful response) to verbal commands, either alone or accompanied by light tactile stimulation. No interventions are required to maintain a patent airway and spontaneous ventilation is adequate. Cardiovascular function is usually maintained.

III. Deep Sedation/Analgesia: A drug-induced depression of consciousness during which patients cannot be easily aroused but respond purposefully following repeated or painful stimulation. The ability to independently maintain ventilatory function may be impaired. Patients may require assistance in maintaining a patent airway and spontaneous ventilation may be inadequate. Therefore, only Anesthesiologists, physicians certified in Critical Care and physicians who are ACLS/PALS certified may perform Deep Sedation. Cardiovascular function is usually maintained.

IV. Anesthesia: Anesthesia consists of general anesthesia and spinal or major regional anesthesia. It does not include local anesthesia. General anesthesia is a drug-induced loss of consciousness during which patients are not arousable, even by painful stimulation. The ability to independently maintain ventilatory function is often impaired. Patients often require assistance in maintaining a patent airway; and positive pressure ventilation may be required because of depressed spontaneous ventilation or drug-induced depression of neuromuscular function. Cardiovascular function may be impaired.

V. Dissociative Sedation: Defined as trance-like cataleptic state induced by an agent such as ketamine. It is characterized by profound analgesia and amnesia, with retention of protective airway reflexes, spontaneous respirations and cardiac stability. After administration of a dissociate agent, the patient passes into a trance. The eyes may remain open but the patient does not respond.

**GENERAL DEFINITION:**

Credentialed indicates that competency is established and documented in the licensed practitioner competency file.

**GENERAL INFORMATION:**

I. Moderate sedation is performed under the direction of a physician credentialed in moderate sedation.

II. Conscious sedation IV medications may be administered by an RN in incremental doses with a physician's order and with the credentialed physician present.

III. Moderate sedation practices and outcomes are evaluated by the appropriate designated personnel.

IV. Documentation is department/unit/service specific and contains all of the required elements.

V. Emergency equipment is immediately assessable to every location where moderate sedation is administered. Opioide and benzodiazepine antagonists, a defibrillator and appropriate equipment for establishing a patent airway and providing positive pressure ventilation with supplemental oxygen is present. Advanced airway equipment and resuscitation medications are immediately available in the crash carts for pediatrics and for adults.

VI. A history and physical is performed by a physician and is in the record prior to the procedure for all patients receiving moderate sedation following #ORG# guidelines.

VII. Peri-procedure assessment is performed and documented including all required elements.

VIII. A "time-out" takes place prior to the procedure in accordance with #ORG# Policy related to Surgical Site Identification.

IX. A malignant hyperthermia cart is present for any patient that receives a medication that could contribute to the onset of malignant hyperthermia. All caregivers are trained in rescue and response for malignant hyperthermia with validated competencies in place.

**PROCEDURES:**

**REFERENCES:**

**FORMS:**

**EQUIPMENT:**

**APPROVALS:**

| NAME | TITLE | DATE |
|------|-------|------|
| #APPROVER# | #APPRTITLE# | #APPRDATE# |

## Policy and Procedures

Policy#: #POLNUM#
Location: #LOC#
Originating Department: #ORIGDEPT#
Effective Date: #EFFDATE#
Expiration Date: #EXPDATE#

## TITLE: NURSING COMPETENCY PLAN

### POLICY STATEMENT:

It is the policy of #ORG# to provide quality nursing care that is individualized by need and provided by the most appropriate member of the health care team (licensed and ancillary staff). #ORG# utilizes a range of methodologies to assess, measure and document, on an on-going basis, the competency of the personnel who comprise the health care team.

### INTENT AND SCOPE:

This policy is intended to establish standardized procedure for competency validation of nursing service personnel. This policy applies to all employees (contract and non-contract) of #ORG#.

### DEFINITIONS:

I. Competency—An employee's ability to demonstrate proficiency of his/her knowledge, critical thinking skills, attitudes, and behaviors in the delivery of care according to his/her job description.

II. General Orientation—A program designed to acquaint the new employee to #ORG# organizational philosophy, policies and procedures, benefits, fire and safety, security and infection control.

III. General Nurse Orientation (GNO)—A program designed to acquaint the nursing employee who is new to #ORG# with an overview of the information critical to the licensed nurse employee of #ORG#.

IV. Department Orientation—A program designed to acquaint the new employee with departmental policies, procedures, and objectives required to successfully complete departmental orientation requirements.

V. Competency Based Orientation Program (CBO)—A program designed to guide personnel in developing and validating competence to perform duties assigned.

VI. Revalidation Process—Annual skill competency assessment defined by each department.

### GENERAL INFORMATION:

I. The departments and service areas of #ORG# designate individuals to monitor the skill level and performance of nursing and all supporting staff.

II. Competency skills for each area have been identified by nursing leadership and approved by nursing management.

III. Skills identified are specific to the growth and developmental needs of the population being served in each area.

IV. Assigned nursing duties are in accordance with the individual's job description, specialized qualifications and competence of the nursing staff available.

V. All new hire employees of #ORG# attend General Orientation within 30 days of employment.

VI. Mandatory training and skills checklists are created for each department.

VII. All new, status change, cross training or transferring employees of the #ORG# attend Departmental Orientation within two (2) weeks of hire, status change, transfer or arrangements for cross training.

VIII. Competency Based Orientation Program (CBO):
   A. All employees of #ORG# (new hire, status change, cross training, or transfer) satisfactorily complete an area specific CBO program within ninety (90) days of hire, status change, cross training request, or transfer.
   B. Competency based skills assessment is satisfactorily completed in collaboration with the orientee, staff Development personnel or designee, assigned preceptor(s) and leadership staff.
   C. Skills may be validated by the manager or designee.
   D. Completion of CBO is documented by area-designated personnel and maintained in the orientee's area personnel file. Appropriate documentation is sent to Human Resources (HR) upon the completion of the requirements of orientation (constitution of appropriate documentation is determined by HR).

IX. Competency Revalidation Process:
   A. Prior to annual Performance Appraisal, the area specific annual Competency Revalidation Process is initiated by area designated personnel.
   B. The area specific Competency Revalidation process is completed by the employee being evaluated and submitted to the evaluator prior to the Performance Appraisal date.
   C. Competency Revalidation includes procedures and skills that are considered:
      1. High risk, high volume;
      2. High risk, low volume;
      3. Performance Improvement/Quality Improvement "fallout"; and,
      4. Individual for employee as well as #ORG# mandatory testing and departmental testing determined for area.
   D. All direct care nursing personnel must attend the annual skills fair.

X. Cross-training;
   A. Before personnel can be cross-trained to another work setting they complete the competency requirements for their primary work setting.
   B. The secondary work setting provides a competency based orientation program for the cross-trainee equal to the program provided primary personnel.
   C. Personnel who are cross training do not function independently until the competency based orientation program for that clinical area is completed. Records of competency validation are maintained in the employee file for the area providing the cross training and the HR employee file.

XI. Agency:
   A. Before beginning work for #ORG#, agency (non-employee) personnel attend a General Orientation conducted by Human Resources. Records of attendance are maintained in #ORG# Staffing Office.
   B. Each department leader is responsible for arranging General Orientation for agency (non-employee) personnel through the HR department as need arises.
   C. The Director of Nursing (DON) is responsible to provide for adequate orientation, supervision and evaluation of the clinical activities of agency (non-employee) nursing personnel occurring within the responsibility of Nursing Services.
   D. Specific Departmental Orientation is provided by and records of it are maintained.
   E. Revalidation of mandatory training, skills assessment and verification of licensure and current American Heart BLS training occur annually. Documentation is maintained in #ORG# administration.
   F. On-going evaluation occurs per #ORG# Policy.

**PROCEDURES:**

**REFERENCES:**

**FORMS:**

**EQUIPMENT:**

**APPROVALS:**

| NAME | TITLE | DATE |
|------|-------|------|
| #APPROVER# | #APPRTITLE# | #APPRDATE# |

**Policy and Procedures**

Policy#:  #POLNUM#
Location:  #LOC#
Originating Department:  #ORIGDEPT#
Effective Date:  #EFFDATE#
Expiration Date:  #EXPDATE#

## TITLE: POINT-OF-CARE TESTING AND PROFICIENCY

### POLICY STATEMENT:

It is the policy of #ORG# that Federal and accreditation requirements for participation in an approved proficiency survey program for procedures performed at the point-of-care are followed.

### INTENT AND SCOPE:

This policy defines a mechanism for the administration of a proficiency survey program for non-Laboratory testing personnel at the point-of-care throughout #ORG#. It applies to all Medical Staff and employees (contract and non-contract) of #ORG#.

### DEFINITIONS:

### GENERAL INFORMATION:

I. A proficiency survey program is a continuous quality improvement resource to check analytic methods, instrumentation, reagent preparation, daily quality control and the competence of personnel responsible for reporting accurate results and completing paperwork for point of care documentation.

II. Proficiency testing (PT) is an external quality control mechanism that involves the analysis of simulated patient specimens with unknown concentrations of specific analytes or organisms provided by an outside laboratory source.

III. Proficiency survey specimens are tested in the same way that patient specimens are tested. Analysts do not discuss, share specimens or results information with any other laboratory until outcome evaluations have been returned from the survey provider.

IV. The senior authority designated for each site location throughout the organization is responsible and accountable for compliant performance of laboratory proficiency testing.

V. The selected contract laboratory provides guidance, staff, and resources to facilitate and assist the employees designated for each site location in proper and compliant management of proficiency testing.

VI. Each patient testing site is enrolled and participates in three (3) testing events per year for all tests performed for which proficiency survey materials are available.

VII. Test results are returned to the proficiency survey program and compared with results from other sites using like instruments and methods.

VIII. Survey scores are returned to the participants. Results of regulated analytes are forwarded to the Center for Medicare and Medicaid Services (CMS).

IX. Each test site must, on a 12-month, rolling calendar basis, achieve satisfactory performance; a score of eighty percent (80%) correct responses on proficiency testing for two (2) consecutive or two (2) out of three (3) survey events for every analyte evaluated.

    A. Determining and returning an incorrect result is considered unsatisfactory performance.

    B. Failure to return the results to the program within the timeframe specified is considered unsatisfactory performance.

    C. Failure to participate in two (2) out of three (3) survey events is considered unsatisfactory performance.

X. Unsatisfactory performance for any reason is addressed through documented remedial action. Documentation is maintained at the test site for two (2) years from the date of participation in the survey event.

XI. Unsatisfactory performance, failure to treat survey specimens in a manner equivalent to patient test samples or referral of proficiency samples to another site for testing results in the imposition of sanctions that may include any or all of the following:

    A. Suspension of a site's privilege to perform patient testing for that failed analyte, specialty or subspecialty;

    B. Limitation, suspension or loss of Medicare and Medicaid monies; and/or

    C. Revocation of the site's CLIA Certificate;

    D. Sanctions for unsatisfactory performance can be withheld if patient testing is suspended voluntarily.

## PROCEDURES:

I. The Contracted Laboratory:

    A. Enrolls each non-laboratory performing point-of-care testing in an approved proficiency survey program for all tests performed for which survey materials are available and that each site is registered to perform.

    B. Verifies proper management of proficiency survey testing events.

    C. Assigns staff to perform the following tasks in coordination with the analysis of proficiency testing materials:

        1. Visits non-laboratory testing sites that have received proficiency testing materials and offer guidance in the proper management of survey materials

        2. Verifies that reports have been completed and signed by the analysts and signed by the site-specific laboratory director prior to submission.

        3. Verifies that returned summary reports are reviewed and signed by the site-specific laboratory director or a technical supervisor from the Contract Laboratory and that corrective action is taken when appropriate.

II. Following receipt of survey materials, the employee designated for each site or designees confirm that:

    A. Survey materials are appropriate or available, i.e., samples are included for analyte testing, sample vials are not broken, etc., and notifies the survey program if sample replacement is deemed necessary;

    B. Photocopies are made of all documentation included with the shipment;

    C. Survey samples are distributed to the specified testing sites for analysis;

    D. Test results are retrieved (in person or via FAX), results are documented on the original survey report form, signed, and the report is photocopied;

    E. Attestation signatures declaring that proficiency testing materials were analyzed using acceptable protocols are obtained from individual(s) performing the testing and the laboratory medical director identified on the site-specific CLIA certificate;

    F. Survey test reports are returned to the program within the timeframe specified on the documents received with the survey;

    G. If a result is graded as unsatisfactory, employee designated for each site or designee makes sure that an investigation is conducted to determine cause of the error, and that any corrective action indicated is taken.

    H. All documentation of all testing activity, including copies of report forms used to record proficiency testing results and the attestation statement provided by the program, signed by the analyst and the Laboratory Director identified on the site's CLIA certificate, is retained for a minimum of two (2) years from the date of the survey event, and is made available upon request by the Department of Health and Human Services or its designated accreditation agent.

III. Following receipt of survey materials, the individuals designated as testing personnel:
   A. Follow directions on the instruction sheet confirm survey samples are tested in the same manner as patient samples and follow standard precautions where appropriate;
   B. Do not send samples to another site for testing;
   C. Do not discuss results with other personnel or test sites until outcome evaluations have been returned from the survey provider;
   D. Sign the attestation stating that steps A-C are true.

**REFERENCES:**

**FORMS:**

**EQUIPMENT:**

**APPROVALS:**

| NAME | TITLE | DATE |
|------|-------|------|
| #APPROVER# | #APPRTITLE# | #APPRDATE# |

**Policy and Procedures**

Policy#:  #POLNUM#
Location:  #LOC#
Originating Department:  #ORIGDEPT#
Effective Date:  #EFFDATE#
Expiration Date:  #EXPDATE#

## TITLE: QUALITY CONTROL PROTOCOL FOR POINT-OF-CARE TESTING

### POLICY STATEMENT:

It is the policy of #ORG# that protocols are developed for Quality Control (QC) testing prior to patient sample analysis and results reporting in accordance with Federal, licensing, and accreditation and #ORG# standards for point-of-care laboratory tests (POCT) and procedures.

### INTENT AND SCOPE:

This policy communicates a standard QC protocol for point-of-care laboratory testing. This policy applies to all Medical Staff and employees (contract and non-contract) of #ORG#.

### DEFINITIONS:

### GENERAL INFORMATION:

- Quality control (QC)—
  A. Is a specific set of procedures and/or materials designed to ensure test systems are performing properly and that results are reliable and accurate; and,
  B. Includes testing samples or materials of known reactivity and/or non-reactivity to assure the test system is operating within acceptable limits and the documentation of those reactions **before** reporting patient test results.
- Testing personnel must be able to perform QC procedures and be aware of the appropriate management of acceptable and unacceptable QC results.
- All test sites must have documented evidence that regulatory and organizational standards for assuring quality laboratory and patient care are met, and personnel performing tests and procedures are competent.

### PROCEDURES:

 I. Before patient specimens are analyzed and resulted, QC control testing is performed according to the specifications established in each written procedure.

 II. The QC is tested according to the following protocol:
  A. Samples that test for an appropriate reactivity or non-reactivity or specific target value are analyzed and meet established criteria for acceptability prior to patient testing and results reporting.
  B. If QC fails acceptability standards, patient samples are not tested until the problem has been identified and resolved according to the procedures established for that system/test.
  C. If QC failure is not resolved, testing is suspended immediately, and subsequent testing is referred to the contract Laboratory for analysis and reporting until the problem is corrected.

D. All successful and unsuccessful QC activity is documented on appropriate logs or held in a database as appropriate to the procedure.

Personnel responsible for quality or performance improvement in the areas/services where POCT is performed review QC activity a minimum of twice weekly and ensure corrective action is taken as needed.

**REFERENCES:**

**FORMS:**

**EQUIPMENT:**

**APPROVALS:**

| NAME | TITLE | DATE |
|------|-------|------|
| #APPROVER# | #APPRTITLE# | #APPRDATE# |

**Policy and Procedures**

Policy#:   #POLNUM#
Location:   #LOC#
Originating Department:   #ORIGDEPT#
Effective Date:   #EFFDATE#
Expiration Date:   #EXPDATE#

## TITLE: RADIATION SAFETY OPERATING AND SAFETY PROCEDURES

### POLICY STATEMENT:

It is the policy of #ORG# to establish and strictly adhere to safety procedures and to maintain records of compliance and controls in accordance with requirements of the (insert state licensure requirement information—see list of state agencies at the end of this policy or visit this website for more information: https://www.asrt.org/content/GovernmentRelations/LegislativeGuidebook/StateRadControlAndLicensureOffices.aspx).

### INTENT AND SCOPE:

This policy is intended to establish standard procedures for radiation safety and to meet regulatory and legal requirements. This policy applies to anyone working in areas having x-ray equipment registered to #ORG#.

### DEFINITIONS:

General Information (information should be inserted throughout this policy to make it pertinent to your state regulations—see list of state organizations at the end of this policy):

I. Operating and Safety procedures are required. This policy establishes procedures that minimize radiation exposure to patients and employees. They are provided to comply with regulations enforced by licensing and regulatory agencies. The regulations require that each x-ray facility be registered with the Office of Radiation Control. The certificate of registration contains conditions and restrictions that apply to the operation of the x-ray machines within #ORG# as well as a listing of the sections of the regulations that apply. These regulations are available for review in the administrator's office.

The regulations require that a Radiation Safety Officer (RSO) be designated. The Medical Director makes this designation. #ORG# RSO has the responsibility and authority for assuring safe radiation practices and serves as the contact person between #ORG# and the Office of Radiation Control. The Radiation Safety Officer may be outsourced through a contracted service. All concerns on radiation safety and compliance are forwarded to #ORG# RSO.

### PROCEDURES:

   I. Operator and Patient Safety
     A.  Credentialing Requirements
       1.  All operators of x-ray machines, including fluoroscopy, must meet all credentialing requirements.
     B.  Personnel Monitoring Requirements
       1.  Any adult receiving a dose from occupational exposure to radiation in excess of 500 millirem in a year must use an individual monitoring device such as a film badge or thermo luminescent dosimeter. Declared pregnant women who receive a dose from occupational exposure to radiation in excess of 50 millirem during the entire pregnancy must also use an individual monitoring device.

    2. Individual monitoring devices must be worn at the unshielded location of the whole body likely to receive the highest exposure. When a protective apron is worn, the location of the individual monitoring device is typically at the neck (collar).

    3. Additional monitoring devices used for monitoring the dose to the embryo/fetus of a declared pregnant woman must be located at the waist under any protective apron being worn by the woman.

    4. The individual monitoring device shall be assigned to and must be worn by only one individual.

    5. If multiple individual monitoring devices are worn by a declared pregnant woman, dose to the embryo/fetus and the occupational dose to the woman shall be determined.

    6. Individual monitoring devices that are not being worn and the control-monitoring device are stored in an area that is away from rooms where radiation machines are in use. This is in/at individual's personal locker/cabinet on site. The control badge is stored in administration.

    7. The Director of Nursing, with the Radiation Safety Officers guidance, is responsible for the occupational dose records and providing/exchanging the individual monitoring devices on the 15th–21st Bi-monthly. The individual monitoring device readings (film badge reports) are located in administration.

    8. Employees working for another agency that receive an occupational dose, should report that dose to #ORG#'s RSO to be included in the manual record of occupational dose.

  C. Use of Protective Devices

    1. Use protective devices, such as lead aprons, gloves and shields, to reduce exposure to radiation and keep radiation exposure As Low As Reasonably Achievable (ALARA). Protective devices must be used or provided in the following situations:

      a. When it is necessary for an individual other than the patient to remain in the room or hold a patient.

      b. When it is necessary to protect other patients who cannot be moved out of the room.

      c. When the gonads are within 5 centimeters of the x-ray beam, shields must be used unless the use of the shield interferes with the diagnostic procedure.

    2. If fluoroscopic procedures are being performed, protective devices (lead drapes, hinged sliding panels) shall be in place. If sterile fields or special procedures prohibit the use of protective devices, all individuals in the fluoroscopic room must wear protective aprons of 0.5 mm lead equivalent.

    3. Protective devices are stored in the x-ray rooms/area.

    4. Protective devices are checked annually for defects, such as holes, cracks, or tears. This check can be done by visually inspecting or feeling the protective devices or may also be done by x-raying these items. A record is kept of this check. If a defect is found at the time of the annual check or on any other occasion, notify the RSO and remove the device from service until it can be repaired or replaced.

  D. Holding of Patients and/or Film

    1. If a patient or film must be supported during a radiation procedure, use a mechanical holding device when circumstances permit. Mechanical devices cannot be routinely used during the following situations in this facility.

      a. Infants and children.

      b. Geriatric Patients

      c. Trauma Patients

    2. If it becomes necessary for an individual to hold a patient or film, the holder should not be pregnant. The holder wears protective devices and keeps out of the direct beam.

  E. Posting Notices, Instructions, and Reports to Workers and Posting a Radiation Area.

    1. Read the "Notice to Employees" sign posted in the x-ray area at each site.

    2. The certificate of registration, operating and safety procedures, and any notices of violations involving radiological working conditions are located in administration. This manual is kept in the x-ray area at each site.

    3. Rights for radiation workers are noted in the policy.

    4. The room(s) in which the x-ray machine(s) is/are located and operated is a radiation area and is restricted. The radiation area is designated by "Caution Radiation Area" signs.

II. Dose to Operators

  A. Occupational dose limits are found in the manual.

  B. If any employee is pregnant or becomes pregnant, she may voluntarily inform the RSO in writing of the pregnancy. If the RSO is informed of the pregnancy, the facility must ensure that the dose to the embryo/fetus does not exceed 0.5 rem (500 mrem) during the entire pregnancy.

C. Radiation Incident or Overexposure
1. Employees who suspect been exposed to an excessive exposure or a radiation incident, immediately notify the RSO.

III. Operation of the X-ray Machine and Film Processing.
A. Ordering of X-ray Exams
1. No x-ray exams are taken unless ordered by a physician.
B. Operator Position During Exposure
1. The operator must be able to continuously view and communicate with the patient.
2. During the exposure, the operator must be positioned so that the operator exposure is As Low As Reasonably Achievable (ALARA) and/or the operator is protected by a lead apron, gloves, or other shielding.
C. Use of a Technique Chart
1. Use of a technique chart aides in reducing the exposure to the operator and patient and it must be used for all exposures. #ORG#'s technique charts are displayed in the vicinity of the control panel of each x-ray machine and may be written or electronically displayed.
D. Restriction and Alignment of the Beam
1. The useful x-ray beam is restricted to the area of clinical interest. Use the centering and beam-limiting devices (collimator) provided on the x-ray machine.
E. Use of Fluoroscopic Machines
1. Reset the five (5) minute cumulative timer device before each fluoroscopic procedure.
2. For mobile fluoroscopy (i.e. C-Arm) units, a 30-centimeter (cm) source-skin distance (SSD) spacer is used.
3. A 20-centimeter SSD spacer may be used for mobile fluoroscopy during surgery procedures only. The following precautionary measures must be used when a 20-centimeter spacer is used: Close observance of fluoro time and technique. Bio-Med technicians restore the 30-Centimeter spacer immediately upon removal of the fluoroscopic machine from the surgery area.
4. Use protective devices during fluoroscopy.
F. Use of Mobile or Portable Machines
1. Mobile x-ray equipment is mounted on a permanent base with wheels and/or casters for moving while completely assembled.
2. Portable x-ray equipment is designed to be hand carried.
3. During the exposure the operator:
   a. must be positioned so that his/her exposure is as low as reasonably achievable; and/or,
   b. must wear lead apron, gloves if necessary, or be protected by other shielding; and,
   c. Should never be in line with direct beam.
4. An individual does not hold the x-ray tube housing during any radiographic exposure.
G. Film Processing
1. Unexposed film is stored in the film storage areas at each site. (Darkroom Cabinets, Film Bins, and in the Supply Storage Room.) The "first in-first out" method is utilized when dispensing film. Expired "out-dated film" is not used for radiographs.
2. Films are developed by the time and temperature recommended by the x-ray film manufacturer. These specifications are posted near the processor at each site.
   a. Check the temperature at the beginning of the workday. Do not process films unless the temperature is as recommended by the manufacturer. Manual processing temperature is checked throughout the workday.
   b. For automatic processors, run blank films through the processor at the beginning of the workday or when ever clean-up films are required.
3. Expiration dates on film and chemicals are checked periodically. New film or chemicals are rotated so the oldest are used first.
4. Do not use film or chemicals after the expiration date.
5. Chemicals are replaced monthly, or no longer than every three months by contracted vendor according to the manufacturer's or chemical supplier's recommended interval.
6. Safe light(s) in the film processing/loading area are provided under these conditions and should not be changed without authorization from the RSO.
   – Filter type: Kodak GBX2 or equivalent.

    –   Bulb wattage: Not more than 15.
    –   Distance from work surfaces: Minimum 4 feet.

If you see light leaks around the doors, ceilings, or other openings in the darkroom, notify the RSO.
Place work orders with Bio-Med or other appropriate repair personnel to correct the problem.

Alternative Processing Systems
Users of daylight processing systems, laser processors, self-processing (Polaroid) film units, or other alternative processing systems follow manufacturer's recommended procedures for image/film processing.

**REFERENCES:**

**FORMS:**

**EQUIPMENT:**

**APPROVALS:**

| NAME | TITLE | DATE |
| --- | --- | --- |
| #APPROVER# | #APPRTITLE# | #APPRDATE# |

# RADIATION CONTROL/STATE LICENSURE OFFICES

## ALABAMA

Kirk Watley, Director
Office of Radiation Control
State Dept. of Public Health
201 Monroe St/PO Box 303017
Montgomery, AL 36130-3017
334-206-5391
kwhatley@adph.state.al.us
Beverly Jo Carswell
X-ray Compliance
State Department of Public Health
PO Box 303017
Montgomery, AL 36130-3017
334-206-5391
bcarswell@adph.state.al.us
www.adph.org
Licensing: does not license personnel

## ALASKA

Clyde Pearce, Chief
Radiologic Health Program
4500 Boniface Parkway
Anchorage, AK 99507-1270
907-334-2107
clyde_pearce@alaska.gov
www.hss.state.ak.us
Licensing: does not license personnel

## ARIZONA

Aubrey Godwin, Director
Arizona Radiation Regulatory Agency
4814 South 40th St.
Phoenix, AZ 85040-2940
602-255-4845 Ext. 222
agodwin@azrra.gov
Licensure: Shanna Farish, Program Manager
Medical Radiologic Technology Board of Examiners
602-255-4845, Ext. 241
sfarish@azrra.gov

## ARKANSAS

Valerie Brown, Agency Program Coordinator
Radiologic Technology Licensure Program
Arkanas Department of Health
Radiation Control Section
4815 W. Markham St., Slot H-30
Little Rock, AR 72205-3867
Phone: 501-661-2301
valerie.brown@arkansas.gov
www.healthyarkansas.com/rtl

## CALIFORNIA

Gary Butner, Branch Chief
Radiologic Health Branch
Division of Food & Radiation Safety
PO Box 997414, MS-7610
Sacramento, CA 95899-7414
916-440-7899
gary.butner@cdph.ca.gov
http://www.cdph.ca.gov

## COLORADO

Joyce Goldsboro, BA, RTR(M)
X-Ray & Mammography Compliance,
Registration & Certification
4300 Cherry Creek Drive South
Denver, CO 80246-1530
303-692-3446
joyce.goldsboro@state.co.us

## CONNECTICUT

Edward Wilds, Ph.D., Director
Dept. of Public Health
Division of Radiation
79 Elm Street
Hartford, CT 06106-5127
860-424-3029
edward.wilds@ct.sgov
http://www.dep.state.ct.us

## DELAWARE

Frieda Fisher-Tyler, Administrator
Office of Radiation Control
Division of Public Health
417 Federal Street
Dover, DE 19903
302-744-4546
frieda.fisher-tyler@state.de.us
www.delaware.gov

## DISTRICT OF COLUMBIA

Gregory B. Talley, Program Manager
Department of Health
HRLA/Radiation Protection Div.
717 14th Street NW, Room 639
Washington, DC 20005
202-724-8800
greg.talley@dc.gov

## FLORIDA

William A. Passetti, Chief
Dept. of Health/Bureau of Radiation Control
4052 Bald Cypress Way, Bin C21
Tallahassee, FL 32399-1741
850-245-4266
bill_passetti@doh.state.fl.us
Licensing: James A. Futch, Administrator
Radiologic Technology Program
4052 Bald Cypress Way
Tallahassee, FL 32399-1741
850-245-4540
james_futch@doh.state.fl.us
www.doh.state.fl.us/environment/radiation

## GEORGIA

Jenella Forrester, Team Leader
Diagnostic Services X-Ray Program
Department of Human Resources
2 Peachtree Street NW, 33rd Floor
Atlanta, GA 30303-3142
404-657-5400
gcforres@dhr.state.ga.us
www.gaepd.org/document/rmprogram1.html
Licensing: No licensing of personnel

## HAWAII

Russell S. Takata, Program Manager
Dept. of Health./Noise & Radiation Branch
591 Ala Moana Blvd.
Honolulu, HI 96813-4921
808-586-4700
russell.takata@doh.hawaii.gov
Licensing: Jeffery M. Eckerd, Supervisor
Radiation Section
Radiological Response/Radiologic
Technology/Mammography
808-586-4700
jeffrey.eckerd@doh.hawaii.gov
www.state.hi.us

## IDAHO

David Eisentrager, Manager
Dept. of Health & Welfare
Idaho Bureau of Laboratories
2220 Old Penitentiary Rd.
Boise, ID 83712-8299
208-334-2235 ext.245
eisentra@dhw.idaho.gov
Licensing: Does not license personnel
www.state.id.us

## ILLINOIS

Joseph Klinger, Asst. Director
IL Emergency Management Agency
Division of Nuclear Safety
1035 Outer Park Dr.
Springfield, IL 62704
217-785-9868
joe.klinger@illinois.gov
Licensing: Steve Collins
Registration and Certification Section
217-785-6982/Fax 217-785-9946
steve.collins@illinois.gov
www.iema.illinois.gov

## INDIANA

John H. Ruyack, Director
State Department of Health
Epidemiology Resource Center/
Indoor and Radiological health
2525 North Shadeland Avenue, E3
Indianapolis, IN 46219
317-351-7190, Ext. 257
Licensing: Darleen Hopper, Coordinator
Operator Certification Program
Medical Radiology Services
2 North Meridan Street, 5F
Indianapolis, IN 46204-3003
317-233-7565
dhopper@isdh.in.gov

## IOWA

Melanie Resmusson, Chief
Bureau of Radiological Health
Lucas State Office Bldg., 5th Fl
321 E. 12th St.
Des Moines, IA 50309-4611
515-281-3478
mrasmuss@idph.state.ia.us
Licensing: Charlene Craig,
Training and Operator Credentialing
515-281-0415
ccraig@idph.state.ia.us
www.state.ia.us

## KANSAS

Tom Conley, Chief
Radiation Section
1000 SW Jackson St, Suite 310
Topeka, KS 66612-1366
785-296-1565
tconley@kdheks.gov

www.kdheks.gov/radiation
Licensing: Kansas Board of Healing Arts
785-296-3680

## KENTUCKY

Dewey Crawford, Manager
Radiation Control Program
Cabinet for Health & Family Services
275 East Main Street, HS1C-A
Frankfort, KY 40621-0001
502-564-3700 Ext. 3695
dewey.crawford@ky.gov
www.chs.ky.gov./publichealth/radiation.htm
Licensing: Vanessa Breeding
Radiation Control Branch
502-564-3700 ext.3693
vanessa.breeding@ky.gov

## LOUISIANA

Jeffery Meyers, Administrator
Emergency & Radiologic Services Div.
PO Box 4312
Baton Rouge, LA 70821
225-219-3041
jeff.meyers@la.gov
http://www.deq.louisiana.gov
Licensing:
Radiological Technologist Board of Examiners
504-838-5231
www.deq.state.la.us/

## MAINE

Jay Hyland Manager
Division of Environmental Health
Radiation Control Program
286 Water Street, 4th Floor
Augusta, ME 04333
Telephone: 207-287-5677
jay.hyland@maine.gov
www.maineradiationcontrol.org
Licensing: Radiological Technology Board
State House Station #35
Augusta, ME 04333-0035
207-624-8603

## MARYLAND

Roland Fletcher, Manager
Radiologic Health Program
Maryland Dept of the Environment
1800 Washington Blvd., Suite 750

Baltimore, MD 21230-1724
410-537-3300
rfletcher@mde.state.md.us
Licensing: Maryland Board of Physicians
410-764-4777
www.mbp.state.md.us

## MASSACHUSETTS

Robert Walker, Director
Radiation Control Program
Department of Public Health
Schrafft Center, Suite 1M2A
529 Main Street
Charlestown, MA 02129
617-242-3035
bob.walker@state.ma.us
www.mass.gov/dph/rcp

## MICHIGAN

Bruce Matkovich
Radiation Safety Section
Div. of Health Facilities & Services
Bureau of Health Systems
MI Dept. of Community Health
PO Box 30664
Lansing, MI 48909
517-241-1993
bmatko@michigan.gov
www.michigan.gov/rss

## MINNESOTA

Dale Dorschner, Manager
Section of Indoor Environments,
& Radiation
Division of Environmental Health
Department of Health
625 Robert Street N.
P.O. Box 64975
St. Paul, MN 55164-0975
651-201-4545
Dale.Dorschner@state.mn.us
www.health.state.mn.us/xray

## MISSISSIPPI

B. J. Smith, Radioactive Materials Director
Division of Radiological Health
State Department of Health
3150 Lawson Street
Jackson, MS 39215-1700
601-987-6893

bjsmith@msdh.state.ms.us
www.health.ms.gov
Licensing: Jimmy Carson
Health Physicist Administrative
X-Ray Branch
Telephone: 601/987-6893
jcarson@msdh.state.ms.us

## MISSOURI

John Langston, Healthcare Regulatory Supervisor
Medical Radiation Control Program
Health Services Regulation
Division of Regulation and Licensure
PO Box 570
Jefferson City, MO 65102-0570
573-751-6083
john.langston@dhss.mo.gov
www.dhss.mo.gov/radprotection/
Licensing: no personnel licensing

## MONTANA

Roy Kemp
Radiological Health Program
MT Dept. of Public Health and Human Services
Licensure Bureau
P. O. Box 202953
Helena, MT 59620-2953
406-444-2868
Licensing: Board of Radiologic Technologists
Montana Dept. of Public Health
3091 South Park 4th Floor
Helena, MT 59620
406-841-2385
www.discoveringmontana.com/dli/bsd

## NEBRASKA

Julia A. Schmitt, Manager
Office of Radiological Health
Dept. of Health & Human Services
P. O. Box 95026
Lincoln, NE 68509-5026
402-471-0528
julia.schmitt@nebraska.gov
www.dhhs.ne.gov/rad

## NEVADA

Karen K. Beckley, M.P.A., M.S.
Program Manager
Radiological Health Section
Bureau of Health Protection Services

Nevada State Health Division
4510 Technology Way, Suite 300
Carson City, NV 89706
775-687-7540
kbeckley@health.nv.gov
www.state.nv.us/health

## NEW HAMPSHIRE

Dennis P. O'Dowd, Administrator
Radiological Health Section
Division of Public Health Srvices
Dept. of Health and Human Services
29 Hazen Drive
Concord, NH 03301-6504
603-271-4585
dodowd@dhhs.state.nh.us
www.dhhs.state.nh.us/dhhs/brh

## NEW JERSEY

Paul Baldauf, Assistant Director
for Radiation Protection Programs &
Release Prevention Element
Dept. of Environmental Protection
P. O. Box 415
Trenton, NJ 08625-0415
609-984-5636
paul.balduaf@dep.state.nj.us
www.state.nj.us/dep/rpp/index.htm
Licensing: Ramona Chambus, Supervisor
Mammography Technologic Certification
609-984-5356
Ramona.chambus@dep.state.nj.us
www.state.nj.us/dep/rpp/index.htm

## NEW MEXICO

Stephen Sanchez, Administrator
Radiologic Technologist Cert. Program
New Mexico Environment Department
1190 St. Francis Drive
Santa Fe, NM 87502-0110
505-476-3264
stephen.sanchez@ state.nm.us
www.nmenv.state.nm.us/nmrcb/home.html

## NEW YORK

Gene Miskin, Director
Office of Radiologic Health
2 Lafayette Street, 11th Floor
New York, NY 10007
212-676-1550

gmiskin@health.nyc.gov
www.health.state.ny.us
Licensing:
NYSDOH, Bureau of Envir. Radiation Protection
547 River St, Room 530
Troy, NY 12180-2216
518-402-7580/Fax 518-402-7575
asa01@health.state.ny.us

## NORTH CAROLINA

Walter Cox, Acting Director
North Carolina Radiation Protection Section
3825 Barrett Drive
Raleigh, NC 27609-7221
919-571-4141, Ext.232
lee.cox@ncdenr.gov
www.ncradiation.net
Licensing: no licensing of personnel

## NORTH DAKOTA

Terry O'Clair, Director
Division of Air Quality
North Dakota Dept. of Health
918 E. Divide Avenue
Bismarck, ND 58501-1947
701-328-5188
toclair@nd.gov
www.ndhealth.gov/aq/rad

## OHIO

Roger E. Owen, Chief
Bureau of Radiation Protection
Ohio Dept. of Health
246 North High Street
Columbus, OH 43215
614-644-2727
robert.owen@odh.ohio.gov
www.odh.ohio.gov
Licensing: James O. Castle, Administrator
Radiologic Technology Section (X-Ray)
614-644-2727
james.castle@odh.ohio.gov
General Inquiries: bradiation@odh.ohio.gov
www.odh.state.oh.us

## OKLAHOMA

Ted Evans, Director
Consumer Protection Services
State Department of Health
1000 Northeast Tenth Street
Oklahoma City, OK 73117-1299

405-271-5243
Licensing: does not license personnel

## OREGON

Terry D. Lindsey, Program Director
Radiation Protection Services
Oregon Health Services,
Department of Human Services
800 NE Oregon Street, Suite 640
Portland, OR 97232-2162
971-673-0499
terry.d.lindsey@state.or.us
www.oregon.gov/DHS/ph/rps/about_us.shtml
Licensing: Oregon Board of Radiologic Technology
503-731-4088 x21

## PENNSYLVANIA

David J. Allard, CHP, Director
Bureau of Radiation Protection
Rachel Carson State Office Bldg.
P.O. Box 8469
Harrisburg, PA 17105-8469
717-787-2480
djallard@state.pa.us
www.dep.state.pa.us
Licensing: Pennsylvania Department of Health
webmaster@health.state.pa.us

## PUERTO RICO

Raul Hernandez, Director
Radiological Health Division
Department of Health
P. O. Box 70184
San Juan, PR 00936-8184
787-274-7802
rhernandez@salud.gov.pr

## RHODE ISLAND

Marie Stoeckel, Chief of Operations
Office of Facilities Regulation
Division of Environmental and Health Sevices
3 Capitol Hill, Room 206
Providence, RI 02908-5097
401-222-4520
marie.stoeckel@health.ri.gov
http://www.state.ri.us
Licensing: Carol Horibin
Radiological Health Specialist
X-Ray Program
401-222-7761

carol.horibin@health.ri.gov
www.health.ri.gov/hsr/professions/rad_tech.php

## SOUTH CAROLINA

Aaron Gantt, Chief
Bureau of Radiological Health
Dept. of Health & Environment Control
2600 Bull Street
Columbia, SC 29201
803-545-4420
ganttaa@dhec.sc.gov
www.scdhec.net Licensing:
South Carolina Radiation Quality Standards Association
803-771-6141/fax 803-771-8048
scrqsa@capconsc.com
www.scrqsa.org

## SOUTH DAKOTA

Bob Stahl, Administrator
Office of Health Care Facilities
Licensure and Certification
615 East 4th St.
Pierre, SD 57501-1700
605-773-3356
bob.stahl@state.sd.us
sd.gov/state_agencies.aspx
Equip. Licensure: Gary Kaus
605-642-6010
gary.kaus@state.sd.us
http://www.state.sd.us
Licensing: does not license personnel

## TENNESSEE

Chuck Johnson, Division of Radiological Health
37911 Middlebrook Pike
Knoxville, TN 37921
Telephone: 865/594-5577
chuck.johnson@state.tn.us
www.state.tn.us/environment/rad/
Licensing:
Tennessee Board of Medical Examiners
Health Related Board
1st Floor, Cordell Hall Bldg.
426 Fifth Ave. N
Nashville, TN 37247-1010
1-800-778-4123

## TEXAS

Richard Ratliff, PE, LMP, Chief
Radiation Program Officer
Bureau of Radiation Control

State Dept of Health Services
PO Box 14937
Austin, TX 78714-9347
512-834-6679
richard.ratliff@dshs.state.tx.us
www.dshs.state.tx.us/radiation
Cathy Fontaine, Mammography Certification
512-834-6688, Ext. 2245
cathy.fontaine@dshs.state.tx.us
Licensing: Pam Kaderka
Texas Department of State Health Services
Medical Radiologic Technologist Certification
   Program
1100 West 49th St.
Austin, TX 78756
512-834-6688
Pam.Kaderka@dshs.state.tx.us
http://www.dshs.state.tx.us/mrt/default.shtm

## UTAH

Dane Finerfrock, Director
Div. of Radiation Control
168 North 1950 West
PO Box 144850
Salt Lake City, UT 84114-4850
801-536-4257
dfinerfrock@utah.gov
www.radiationcontrol.utah.gov
Richard Sanborn, Health Physicist
(Mammography Contact)
801-536-4268
rsanborn@utah.gov
Licensing: Occupational and Prof. Licensure
PO Box 45805
Salt Lake City, UT 84114-6741
801-530-6403

## VERMONT

William Irwin, Sc.D., CHP, Chief
Office of Radiologic Health
Department of Health
108 Cherry Street
PO Box 70
Burlington, VT 05402
802-865-7730
wirwin@vdh.state.vt.us
healthvermont.gov/enviro/rad/rad_health.aspx
Carla White, Sr. Radiological Health Specialist
Office of Radiological Health
(Mammography Contact)
802-862-8171
cwhite@vdh.state.vt.us

## VIRGINIA

Leslie P. Foldesi, CHP, Director
Division of Radiological Health
Department of Health
James Madison Bldg.
109 Governor Street, Room 730
Richmond, VA 23219
804-864-8151
les.foldesi@vdh.virginia.gov
www.vdh.virginia.gov
Stan Orchel, Assistant Director
X-Ray Registration and Certification
(Mammography Contact)
804-864-8170
stan.orchel@vdh.virginia.gov
Licensing:
Board of Medicine for Radiologic Technology
6603 West Broad Street, 5th Floor
Richmond, VA 23230-1712
804-662-9908
www.dhp.state.va.us/medicine/default.htm

## WASHINGTON

Gary L. Robertson, Director
Office of Radiation Protection
Department of Health
PO Box 47827
Olympia, WA 98504-7827
360-236-3210
Gary.robertson@doh.wa.gov
www.doh.wa.gov/ehp/rp
Scott Mantyla, X-Ray Section
(Mammography Contact)
360-236-3232
scott.mantyla@doh.wa.gov

## WEST VIRGINIA

Randy Curtis, Director
Radiological Health Program
Office of Environmental

Health Services
DHHR Bureau for Public Health
1 Davis Square, Suite 200
Charleston, WV 25301-1798
304-558-6721
randy.c.curtis@wv.gov
www.wvdhhr.org/rtia
Licensing: Grady M. Bowyer, R.T.(R), Executive
Director
WV Radiologic Technology Board of Examiners
304-546-4642
gradymbowyer@suddenlink.net
www.wvrtboard.org

## WISCONSIN

Paul Schmidt, Manager
Radiation Protection Section
Dept. of Health and Family Services
PO Box 2659
Madison, WI 53701-2659
608-267-4792
paul.schmidt@wisconsin.gov
dhfs.wisconsin.gov/dph_beh/RadiatioP/Index.htm

## WYOMING

Dewey Long, MS
Mammography Program
Department of Health
Office of Healthcare Licensing
6101 Yellowstone Road, Suite #400
Cheyenne, WY 82002
307-777-5244
dewey.long@health.wyo.gov
Licensing: Wyoming Board of Radiologic
Technologist Examiners
First Bank Plaza, Suite 201
2020 Carey Ave.
Cheyenne, WY 82002
307-777-3507
plboards.state.wy.us/radiology

**Policy and Procedures**

Policy#:  #POLNUM#
Location:  #LOC#
Originating Department:  #ORIGDEPT#
Effective Date:  #EFFDATE#
Expiration Date:  #EXPDATE#

## TITLE: REPORTING PANIC AND URGENT LABORATORY RESULTS

### POLICY STATEMENT:

It is the policy of #ORG# to follows a protocol for prompt reporting of "Panic Results" and "Urgent Results" in accordance with regulatory standards.

### INTENT AND SCOPE:

This policy establishes and communicates a mechanism to ensure critical laboratory test results are reported to appropriate personnel in a timely manner to provide quality patient care. This policy applies to all Medical Staff and employees (contract and non-contract) of #ORG#.

### DEFINITIONS:

### GENERAL INFORMATION:

   I. A "panic result" is a laboratory test result that is potentially life threatening, and requires immediate action by designated personnel to report the value to the appropriate practitioner or patient care personnel.

   II. An "urgent result" is a laboratory test result that is significantly abnormal and requires direct attention by designated personnel to report the value to the appropriate practitioner or patient care personnel.

   III. "Timely" is defined as timeframe from receipt of the critical result report from the laboratory to notification of the physician or mid-level practitioner.
     A.  Panic Result:
       1.  Immediate patient assessment, report to physician within fifteen (15) minutes.
       2.  Outpatient: attempts to contact the patient within thirty (30) minutes.
     B.  Urgent Result:
       1.  Patient assessment dependent and report to physician within two (2) hours.
       2.  Outpatient—site notification the morning of the next business day.

   IV. Panic Result and Urgent Result reporting is managed in accordance with regulatory guidelines established by the College of American Pathologists (CAP) and state licensing requirements, along with CMS requirements.
     A.  Documented evidence of communication is required to indicate that appropriate clinical personnel are notified of all panic values and urgent values for laboratory tests when results indicate that a patient's life may be in immediate danger.
     B.  Panic result and urgent result notification is accomplished as defined in the PROCEDURE below.
     C.  Panic results and urgent results are reported to licensed nursing staff only. Acknowledgment of panic result and urgent result receipt is required. The person receiving the result must "read back" or repeat the reported information as validation that the information was understood as delivered.

V. Reportable panic result and urgent result ranges are defined by the contract Laboratory Medical staff in conjunction with the Medical Director and are subject to revision as deemed necessary.

VI. Point-of-Care laboratory testing results are managed immediately by the individual performing the analysis.

**PROCEDURES:**

I. The Technologist/Technician who performs the analysis reports panic results and urgent results to appropriate patient care location.
  A. All panic results and urgent results are immediately phoned to the originating patient care area on a 24/7/365 basis.
  B. The reporting technologist verifies that the individual receiving the panic result and/or urgent result report is a licensed nurse.
  C. Recipients of panic result and urgent result reports proceed according to policy.
  D. If a dilution is required to measure a therapeutic drug or a drug of abuse, the Technologist performing the test will notify patient care personnel in the appropriate location that results are above instrument range, require dilution and reanalysis, and will likely produce a critically high result. Final results are called upon test completion.

II. Panic result and urgent result communications, including failed attempts, are documented by the Technologist and communicated to the facility.

III. The risk management occurrence system will be used to report any issues in the critical result notification process.

**REFERENCES:**

**FORMS:**

**EQUIPMENT:**

**APPROVALS:**

| NAME | TITLE | DATE |
|------|-------|------|
| #APPROVER# | #APPRTITLE# | #APPRDATE# |

**Policy and Procedures**

Policy#:  #POLNUM#
Location:  #LOC#
Originating Department:  #ORIGDEPT#
Effective Date:  #EFFDATE#
Expiration Date:  #EXPDATE#

## TITLE: RISK MANAGEMENT: ANALYSIS AND REPORTING
## SENTINEL EVENT, ADVERSE EVENTS, NEAR MISSES

### POLICY STATEMENT:

It is the policy of #ORG# to identify, review and manage sentinel events and other reportable adverse events defined by state and federal law and near misses.

### INTENT AND SCOPE:

The policy is intended to identify, review and manage reportable sentinel events, other reportable adverse events defined by state and federal law and near misses. This policy applies to all Medical staff and employees (contract and non-contract) of #ORG#.

### DEFINITIONS:

I. "Near Miss" is defined as any process variation which did not affect the outcome, but for which a recurrence carries a significant chance of a serious adverse outcome. Such a near miss falls within the scope of the definition of a sentinel event, but outside the scope of those sentinel events that are subject to review by an accrediting body under its Sentinel Event policy.

II. "Serious Injury" is defined as an unanticipated death, loss of limbs, loss of organs, loss of bodily function, or severe central nervous system impairment.

III. "Root Cause Analysis" (RCA) is a process for identifying the basic or causal factors that underlies variation in performance, including the occurrence or possible occurrence of a sentinel event. A root cause analysis focuses primarily on systems and processes, not individual performance. It progresses from special causes in clinical processes to common causes in organizational systems that would tend to decrease the likelihood of such events in the future, or determines, after analysis, that no such improvement opportunities exist.

IV. "Sentinel Event" is an unexpected occurrence involving death or serious physical or psychological injury or risk thereof. The facility is responsible for conducting a root cause analysis and corrective action plan. The event has resulted in an unanticipated death or major permanent loss of function, not related to the natural course of a patient's illness or underlying conditions;

V. "Specific Adverse Events" is an unexpected occurrence involving death or serious physical or psychological injury or risk thereof. The facility is responsible for conducting a root cause analysis and corrective action plan. The following Adverse Events are subject to annual reporting to #STATE# Department of State Health Service

VI. Serious errors track medical errors and adverse patient events, analyze their causes, and implement preventive actions and mechanisms that include feedback and learning throughout the facility. The facility will take actions aimed at performance improvement and, after implementing those actions; the facility must measure its success, and track performance to ensure that improvements are sustained.

VII. National Quality Forum (NQF) "Never Events" is adverse events that are serious, largely preventable, and of concern to both the public and healthcare providers for the purpose of public accountability

## GENERAL INFORMATION:

I. Individuals who identify and report quality/safety hazards, near misses, or adverse occurrences are not subject to retaliatory actions for reporting the concern or event. Violations of policy/procedure or unsafe work practices may be addressed per the behavior and work standard policy and or the appropriate peer review process

II. Events identified in this policy and the reports thereof are confidential. Thus these reports may not be copied. Any reports are generated at the direction of the Quality and Safety Committee. The report is intended for the purposes of quality improvement activities.

III. Disclosure of outcomes of care including unanticipated outcomes or medical errors that cause significant harm or injury to the patient is made to the patient in accordance with #ORG# Policy. If the patient is deemed incapable of understanding a discussion of this nature, then in accordance with #ORG# Policy, the surrogate decision maker substitutes for the patient.

IV. The responsible caregiver(s) objectively documents the actual event and patient-related follow-up and assessment in the medical record. Do not document the completion of the occurrence report in the medical record or the investigation thereof.

## PROCEDURES:

I. When a physician or staff member witnesses or identifies that a near miss, sentinel event, and other adverse events not related to a Significant Medication Error and/or Significant Adverse Drug Reaction event has occurred.
   A. Immediately and simultaneously notify the Director of Nursing.
   B. Follow #ORG# policy regarding Risk Variance Reports.

II. When a physician or staff member witnesses or identifies that a Significant Medication Error and/or Significant Adverse Drug Reaction event has occurred.
   A. Immediately and simultaneously notify the Director of Nursing and/or Medical Director.

III. The Quality/Risk Officer initiates an investigation of the event as notice is received, with a report to the insurance carrier.

IV. The Quality/Risk Officer contacts the Quality and Safety Committee Chair and Administrator with the initial investigation.

V. If during the course of an investigation the risk review committee determines that a root cause analysis is to be performed, then a team is assigned to perform a root cause analysis.
   A. Multidisciplinary representatives from the area affected in the event may be involved in the analysis. These representatives may include, but are not limited to, physicians, nurses, ancillary services, and administrators, and are considered to be a medical peer review committee.
   B. The Committee uses the approved RCA process to determine what areas need detailed inquiry when conducting an RCA.
   C. A root cause analysis is completed within 45 (forty-five) days of becoming aware of the event.
   D. The team is empowered to redesign systems or processes as necessary to eliminate the root cause. This may involve change in training, policies or procedures, forms, equipment, etc. The team can design and implement changes prior to completing the root cause. Intermediate changes may be necessary to reduce immediate risk.
   E. Findings and recommendations from the root cause analysis are reported to the QSC, the Medical Executive Committee and the Board of Managers.

VI. Upon completion of the RCA, the Committee determines if the event meets the Sentinel Event definition.

VII. Events identified as sentinel events by the accrediting body may be self-reported as per the requirement for self-reporting sentinel events.

VIII. Events identified as reportable by applicable laws are reported accordingly.

**REFERENCES:**

**FORMS:**

**EQUIPMENT:**

**APPROVALS:**

| NAME | TITLE | DATE |
|------|-------|------|
| #APPROVER# | #APPRTITLE# | #APPRDATE# |

**Policy and Procedures**

Policy#:  #POLNUM#
Location:  #LOC#
Originating Department:  #ORIGDEPT#
Effective Date:  #EFFDATE#
Expiration Date:  #EXPDATE#

## TITLE: SAFE MEDICAL DEVICE ACT REPORTING

### POLICY STATEMENT:

It is the policy of #ORG# to promote a safe environment for all patients, employees and visitors. While the task for reporting medical device failures/damage is the responsibility of each #ORG# employee, the appointed Safety Officer and the Quality and Safety Committee function as the primary resources to fulfill the reporting requirements of the Food and Drug Administration (FDA).

### INTENT AND SCOPE:

The policy is intended to establish and communicate a standard approach to the reporting of any medical device that caused a death or serious injury. This policy applies to Medical staff and employees (contract and non-contract) of #ORG#.

## DEFINITIONS:

I. Medical device—anything used in the treatment or diagnosis for health related purposes that are not a drug. Medical devices may include but are not limited to equipment, implants, disposable, and radioactive contrast media.

II. Serious injury/illness—an injury or illness that is either 1) life threatening, 2) results in permanent impairment of a body function or permanent damage to a body structure, or 3) necessitates immediate treatment to prevent permanent impairment of a body function or permanent damage to a body structure.

### GENERAL INFORMATION:

I. The United States Congress enacted the Safe Medical Devices Act (SMDA) (Pub.L.101-629) in November 1991. The SMDA has authorized the FDA to require reporting of any device-related deaths or serious injury beginning July 31, 1996 to the FDA and the manufacturer within ten (10) working days.

### PROCEDURES:

I. In the event that the device is thought to have caused a death or serious injury to a patient, the user/operator performs the following after notifying the appropriate patient care staff.
   A. Tags the defective device and discontinues its use. The subject device causing injury is delivered only to the Safety Officer.
   B. Reports the event to their immediate supervisor as soon as practical, including the Safety Officer in the reporting process.
   C. Removes the device and identifies any accessories or ancillary devices in use at the time of the incident.
   D. Reports to the Biomedical Engineering prior to the end of the working shift.
   E. Notes all pertinent conditions existing at the time to include control settings, ancillary devices in use, condition of the patient, previous indications of problems, etc. on the #ORG# Variance Form.

II. The supervisors perform the following;
    A.  Notification of their superior immediately.
    B.  Forwards the Variance report to the Safety Officer.

III. The Safety Officer provides a report to the Quality and Safety Committee at the next scheduled meeting. In the meantime a Root Cause Analysis is completed per risk management protocol and policy.

IV. Biomedical Engineering, upon direction of the Safety Officer:
    A.  Reports all required events to the manufacturer and/or FDA within ten (10) working days.
    B.  Completes and forwards the semi-annual report to the FDA.
    C.  Maintains all files for a twenty-four (24) month period.
    D.  Reviews the event and reports findings to the Quality and Safety Committee.

**REFERENCES:**

**FORMS:**

**EQUIPMENT:**

**APPROVALS:**

| NAME | TITLE | DATE |
|------|-------|------|
| #APPROVER# | #APPRTITLE# | #APPRDATE# |

**Policy and Procedures**

Policy#: #POLNUM#
Location: #LOC#
Originating Department: #ORIGDEPT#
Effective Date: #EFFDATE#
Expiration Date: #EXPDATE#

## TITLE: SAFE PRACTICES FOR MOVING PATIENTS

### POLICY STATEMENT:

It is the policy of #ORG# to provide a patient lifting, handling and movement program that incorporates safety equipment, proper techniques, safer working conditions, and safety training to employees which includes procedures for reporting workplace hazards and injuries.

### INTENT AND SCOPE:

The policy is intended to provide guidelines for safe patient lifting, handling, and movement that comply with the applicable federal and state laws, regulations and guidelines regarding workplace safety and ergonomics and affects all Volunteers, Students, Medical Staff and employees (contract and non-contract) of #ORG#.

### DEFINITIONS:

I. Hazardous Work Conditions: The potential for harm or damage to people, property, or the work environment. Hazards are classified as physical, chemical, psychological or mechanical.

II. Lifting, Handling and Movement Tasks: Patient related tasks that pose a high risk of musculoskeletal injury to patients or staff performing the tasks. These include but are not limited to transferring, lifting, repositioning, pushing, pulling and/or carrying tasks, bending or stooping, bathing patients, making occupied beds, dressing patients, turning patients in bed, ambulating patients, and tasks with long durations.

III. Manual Lifting: Lifting, transferring, repositioning and moving patients using the employee's body strength without the use of lifting equipment/aids.

IV. Mechanical Lifting Equipment: Equipment used to lift, transfer, reposition and move patients.

V. Patient Handling Aids: Equipment used to assist in the lift or transfer process, such as gait belts, stand assist aids, sliding boards, etc.

### GENERAL INFORMATION:

It is the duty of #ORG# staff to take reasonable care of their own health and safety, as well as that of their co-workers and their patients during patient handling activities by following this policy. #ORG# staff is expected to make safe lifting, handling, and movement of patients a critical part of their job.

### PROCEDURES:

I. Staff analyzes the risk of injury to both patients and staff posed by the patient handling needs of the patient populations served by #ORG# and the physical environment in which patient handling and movement occurs.

II. Staff avoids hazardous patient handling and movement tasks whenever possible. If unavoidable such as a medical emergency, assess and seek help if possible.

III. Use mechanical lifting devices and other approved patient handling aids for high-risk patient handling and movement tasks except when absolutely necessary such as in a medical emergency.

IV. Use mechanical lifting devices and other approved patient handling and movement aids in accordance with instructions and training.

V. Staff completes education on safe patient handling and movement initially, annually, and as needed.

VI. Mechanical lifting devices and other equipment/aids are accessible to staff and in proper working order.

VII. Staff report safety concerns and request assistance through the appropriate chain of command in order to control risks of injury to patients and staff during patient handling.

VIII. Consideration of the feasibility of incorporating patient handling equipment or the physical space and construction design needed to incorporate that equipment at a later date is given when developing plans for constructing or remodeling facilities.

IX. Staff reports all patient/visitor or staff injuries from patient handling/movement according to #ORG# policy.

X. The process of Safe Harbor is in place and may be used by the nurse to refuse to perform or be involved in patient handling or movement that the nurse believes in good faith expose a patient or a nurse to an unacceptable risk of injury.

XI. An annual report on activities related to safe patient handling and analysis of strategies to control risk of injury is submitted to the Quality and Safety Committee on activities related to safe patient handling and movement.

**REFERENCES:**

**FORMS:**

**EQUIPMENT:**

**APPROVALS:**

| NAME | TITLE | DATE |
|------|-------|------|
| #APPROVER# | #APPRTITLE# | #APPRDATE# |

**Policy and Procedures**

Policy#:  #POLNUM#
Location:  #LOC#
Originating Department:  #ORIGDEPT#
Effective Date:  #EFFDATE#
Expiration Date:  #EXPDATE#

## TITLE: SAFETY CONSIDERATION FOR INFANTS AND PEDIATRIC PATIENTS

### POLICY STATEMENT:

It is the policy of #ORG# that infant and pediatric safety measures are followed.

### INTENT AND SCOPE:

The policy is intended to provide safety guidelines for infants and pediatric patients. This policy applies to all students, volunteers, Medical staff and employees (contract and non-contract) of #ORG#.

### DEFINITIONS:

### GENERAL INFORMATION:

Staff members are trained in the care of infants and children in the patient care areas. Infants are not left unattended by parents in non-patient care areas. Siblings are not left unattended in any care setting. Code Pink drills should be performed at least once a year if infants and pediatric patients are cared for in the facility.

### PROCEDURES:

   I. Basic facility needs and safety practices for infants and children consist of:
     A.  Covered electrical outlets, childproof window locks and sharp edges are padded in patient care areas.
     B.  Furniture meets Consumer Product Safety Commission Standards.
     C.  Parents or clinic staff is in attendance for infants and children in an exam room.
     D.  Staff is never to turn their backs to an infant or small child on a high surface. Maintain hand contact on the child's back or abdomen to protect from rolling, crawling or jumping off of all high surfaces.
     E.  Blind cords are secured to wall to avoid strangulation.
     F.  Communal toys in waiting rooms and patient care areas are discouraged due to infection control concerns.
     G.  Single or double occupancy rooms, which are large enough to accommodate parents who wish to stay with their children.
     H.  Patient room configuration and bed positioning that allows convenient observation and supervision of patients by nursing staff, if a parent is not available.
     I.  Cribs are not placed within reach of heating units, dangling cords, or other dangerous objects
     J.  Non-slip easily maintained floor surfaces.
     K.  Age appropriate equipment, including cribs with safe overhead covers and beds with covered electrical controls.
     L.  Separate treatment room for infant and toddler procedures.
     M.  Resuscitation cart containing pediatric specific supplies.
     N.  Bed checks are conducted every shift.
     O.  Crib is positioned furthest from the door.
     P.  Infant warmers to have all four sides up when not performing procedures.

Q.  Cribs to have side rails up at all times.
R.  Staff is in attendance of an infant/child; on a stretcher, exam or operating room table, in a warmer with sides down or crib with rails down. Staff maintains one hand on abdomen or chest to prevent rolling, crawling, or jumping from the surface.
S.  Infants are held while bottle feeding. Only plastic bottles are given to children who hold their own bottles.
T.  Infants are positioned in a supine position on a firm mattress without any potentially suffocating bedding.
U.  Pacifiers are not artificially secured to an infant or an item in the infant's bed.
V.  Co-bedding of infants with a parent or guardian is not permitted.
W.  Infants and children are securely strapped in place in infant seats, feeding chairs, strollers or anti-reflux wedges.
X.  Bath water is checked prior to bathing. Bath water is lukewarm, not hot to the touch. Infants are not placed under running water when washing their head.
Y.  Infants are discharged with an appropriate car seat.
Z.  Children discharged post procedure that are recovering from anesthesia:
    1.  Must be able to hold head up, not compromising airway.
    2.  Must have two people in the car, one to monitor the head and airway, while driving home.


**REFERENCES:**


**FORMS:**


**EQUIPMENT:**


**APPROVALS:**

| NAME | TITLE | DATE |
|---|---|---|
| #APPROVER# | #APPRTITLE# | #APPRDATE# |

**Policy and Procedures**

Policy#: #POLNUM#
Location: #LOC#
Originating Department: #ORIGDEPT#
Effective Date: #EFFDATE#
Expiration Date: #EXPDATE#

## TITLE: SAFETY MANAGEMENT PLAN

### POLICY STATEMENT:

It is the policy of #ORG# to design, implement and monitor a safety program to reduce the risk of injury to patients, staff and visitors.

### INTENT AND SCOPE:

This policy is intended to meet legal or regulatory requirements in accordance with federal, state, local, and regulatory agencies. The scope of #ORG#'s Safety Management Plan is to provide to our patients, personnel and visitors a physical environment free of hazards and to manage activities proactively through risk assessments to reduce the risk of injuries. This policy applies to #ORG# employees (contract and non-contract), medical staff and volunteers.

### OBJECTIVES:

1. Evaluate the effectiveness of all safety processes and to implement improvement initiatives when opportunities are identified.
2. Appoint a safety officer for the facility that oversees development and implementation of the safety program.
3. Develop and implement a safety communication program.
4. Educate all staff, physicians, contract workers and other stakeholders on the safety program.

### DEFINITIONS:

### GENERAL INFORMATION:

There are inherent safety risks associated with providing services for patients, the performance of daily activities by staff, and the physical environment in which services occur. This policy plans and implements procedure to minimize risks.

### PROCEDURES:

I. #ORG# plans activities to minimize risks in the Environment of Care.
  A. The Quality and Safety Committee is responsible for developing, implementing, and monitoring the safety program for #ORG#. The committee:
    1. Complies with safety-related regulatory agency standards and State and Federal laws.
    2. Enforces and monitors current work practices and work conditions for patient, staff, and visitor safety.
    3. Provides safety education to all employees of #ORG#.
    4. Develops performance indicators to identify opportunities for improvement to minimize risks and improve safety performance.
    5. Monitors and evaluates the effectiveness of the safety program.
  B. The Board of Managers appoints the Safety Officer. The Safety Officer is a member of the Quality and Safety Committee. He/she helps the chairperson prepare the agenda and ensures that all safety reporting

is complete. The Safety Officer also identifies safety management issues and coordinates ongoing organization-wide collection of information on occupational hazards and safety practices to identify safety and security issues to be addressed by the Quality and Safety Committee.

C. The Quality and Safety Committee Chairperson appoints members to the committee. Members are chosen to provide a wide cross-section of professions and expertise within #ORG#, and include administrative, clinical and support personnel and others as necessary.

D. Every member of the Quality and Safety Committee: alerts appropriate individuals on situations requiring corrective action; sets a standard for safety awareness; and, encourages other employees to be concerned with good safety practices.

E. The Safety Officer has the authority to take immediate action when conditions or practices exist that could cause personal injury to individuals or damage to equipment or buildings or could reasonably be expected to result in death or serious physical harm before such conditions can be eliminated.

F. The Safety Officer informs the affected department and staff of the hazards and confines conditions or suspends practices until #ORG# or Department implements measures to reduce or eliminate hazards.

G. Quality and Safety Committee members have the authority to enter without delay and at reasonable times any #ORG# facility, department, or area workplace to inspect and investigate conditions, structures, equipment, to question staff (contract or non-contract), and review records or documents which are directly related to and for the purpose of inspection.

H. #ORG# creates and maintains a safe environment and ensures the following:

1. #ORG# has a written and executed Safety Management Plan which describes the processes #ORG# implements to manage the environmental safety of patients, staff and visitors.

2. #ORG# has a written and executed Secure Environment Management Plan which identifies how #ORG# establishes and maintains security issues concerning patients, visitors, personnel, and property.

3. #ORG# has a written and executed Hazardous Materials and Waste Management Plan that describes the processes #ORG# implements to manage hazardous materials and waste.

4. #ORG# has a written and executed Fire Safety Management Plan describing the processes #ORG# implements to manage fire safety.

5. #ORG# has a written and executed Medical Equipment Management plan that describes the processes #ORG# implements to manage medical equipment.

6. #ORG# has a written and executed Utilities Management Plan that describes the processes #ORG# implements to manage utilities.

I. Risk Assessments

1. #ORG# conducts an ongoing proactive risk assessment to evaluate the potential of adverse impacts of buildings, grounds, equipment, occupants and internal physical systems on the safety and health of patients, staff and other visitors. The goal of performing risk assessments is to reduce the likelihood of future incidents or other negative experiences that have the potential to result in an injury, an accident, or other loss to patients, staff or other assets.

2. Risks of safety related incidents are reduced by proactively evaluating systems and making necessary changes through the Quality and Safety Committee, performance improvement, Administration and departmental participation.

J. Action Plans to Reduce and/or Eliminate Risk

1. #ORG# uses the risks and hazards identified to select and implement changes in procedures and controls by:

   a. Creating new or revising safety polices and procedures.
   b. Identifying new environmental round items for the areas affected.
   c. Improving safety orientation and education programs.
   d. Helping define safety performance monitoring and indicators.

2. Maintaining Grounds and Equipment

3. Plant Operations is responsible for scheduling, managing and performing maintenance of facility grounds and external equipment. Plant Operations staff make regular rounds of various areas to observe and correct the current condition and safety of facility grounds and external equipment.

4. Facility grounds include lawns, shrubs and trees, sidewalks, roadways, parking lots, lighting, signage, and fences. Some external equipment such as oxygen storage facility has established protocols for inspection, testing, or preventive maintenance.

K.   Creating a Secure Environment
1.   #ORG# has approved and executed the Secure Environment Management Plan. The plan identifies how #ORG# establishes and maintains a security program for patients, visitors, personnel, and property.
2.   #ORG# controls access to and from areas it identifies as security sensitive.
L.   Responding to a Security Incident
1.   The Secure Environment Management Plan identifies how #ORG# establishes and maintains a security program for patients, visitors, personnel, and property.
2.   Code Pink policy is in place to provide guidance should an infant or pediatric abduction occur.
3.   Drills covering abduction, security events and safety issues throughout the year.
M.   Responding to Product Notices and Recalls
1.   #ORG# monitors regular and critical responses to product safety recalls made by outside organizations and internal department representatives.
2.   The Risk Officer manages the process, receiving reports from manufacturers and vendors and distributing the information to those departments using or managing the products.

**REFERENCES:**

**FORMS:**

**EQUIPMENT:**

**APPROVALS:**

| NAME | TITLE | DATE |
|------|-------|------|
| #APPROVER# | #APPRTITLE# | #APPRDATE# |

**Policy and Procedures**

Policy#:  #POLNUM#
Location:  #LOC#
Originating Department:  #ORIGDEPT#
Effective Date:  #EFFDATE#
Expiration Date:  #EXPDATE#

## TITLE: SCOPE OF CARE AND SERVICES

### POLICY STATEMENT:

#ORG# operates under a formally articulated Scope of Care and Service that is designed to provide the population of #ORG# with health care that is of the highest possible quality in a manner that is cost effective.

## INTENT AND SCOPE:

The intent of this policy is to establish and communicate the process by which new services are added to, or changes to existing services are made in, the Scope of Care and Services of #ORG#. All Medical Staff associated with #ORG# and all employees and contracted services of #ORG# are affected by this policy.

## DEFINITIONS:

### GENERAL INFORMATION:

I. The clinical leadership of #ORG# will:
   A. Evaluate each change in the Scope of Care and Service to allow adequate clinical and administrative evaluation of medical necessity versus total costs, alternatives and need to purchase services to be provided by #ORG#.
   B. Promote value in our services.
   C. Provide the most appropriate means of care, given limited resources to provide care.
   D. Assure that individuals requesting a service are privileged to provide that service.
   E. Carefully analyze each change in service provided by #ORG# if it involves a change in a Medical Standard of Care, a network process or procedure, personnel or equipment.
   F. Assure the coordination between functional units and Physician Services regarding privileging of providers for services to be provided.
   G. Allow some mechanism of resource management control.
   H. To determine merit of purchasing a service or procedure through an outside service versus provision within #ORG#.

II. Any surgery, procedure, or process of care that requires new processes, equipment, floor space, and/or personnel is considered to be a change in the Scope of Care and Service if it is a change from the present published Scope of Care and Service Plan that is approved each year by the Medical Executive Committee and Board of Managers.

III. All completed requests for new, additional or changes in Scope of Care and Service are reviewed by clinical leadership.

IV. There are four areas that must mesh in order for a Scope of Care and Service change to be accurate, thus implemented: competency of the staff, clinical privileges for physicians/practitioners, equipment availability, and instrument availability.

V. If approved, the Scope of Care and Service Plan is changed and presented for approval by the MEC and BOM.

**PROCEDURES:**

**REFERENCES:**

**FORMS:**

**EQUIPMENT:**

**APPROVALS:**

| NAME | TITLE | DATE |
|---|---|---|
| #APPROVER# | #APPRTITLE# | #APPRDATE# |

**Policy and Procedures**

Policy#:  #POLNUM#
Location:  #LOC#
Originating Department:  #ORIGDEPT#
Effective Date:  #EFFDATE#
Expiration Date:  #EXPDATE#

## TITLE: SECURE ENVIRONMENT MANAGEMENT PLAN

### POLICY STATEMENT:

It is the policy of #ORG# to have a Secure Environment Plan to provide a programmatic framework that reduces risk to #ORG#. The plan includes processes that are designed to evaluate hazards that may adversely affect the life or health of patients, staff, and visitors while in the course of its mission of providing a safe, secure, and therapeutic environment.

### INTENT AND SCOPE:

This policy is intended to meet legal or regulatory requirements. #ORG#'s Quality and Safety Committee (QSC) is responsible for overseeing the security of the facility and this committee consists of a cross representation of #ORG#'s staff. The QSC monitors training and competence of staff and assesses conditions of the physical plant, grounds, and equipment through building inspections, environmental rounds, security inspections, and various performance improvement initiatives. Through review of reliable information, administration is able to make the best decisions regarding security concerns and to evaluate security performance related to key issues with opportunities for improvement. The QSC monitors and evaluates all safety issues. It takes action and makes recommendations to #ORG# leadership, including the organization's administrator, who is a member of the Board of Managers (BOM). This policy applies to all employees (contract and non-contract) of #ORG#.

### OBJECTIVES:

1. How to identify and address security issues the impact patients, visitors, employees, property and physicians/practitioners.
2. Appointment of employees responsible for security.
3. Reporting and investigating security incidents.
4. Controlling access to sensitive and high risk areas.
5. Education and training of staff, physicians, practitioners and contract workers.
6. Evaluating the plan at least annually.

### DEFINITIONS:

### GENERAL INFORMATION:

I. Uniformed, visible security presence in #ORG# leads to a reduction in security incidents and to a feeling of greater security by patients, staff, and visitors. Further, the professional investigation of all criminal offenses effectively reduces the number of repeat criminals within #ORG#.

II. Ongoing assessments of risks and ongoing programs of patrolling to identify risks and problems permit timely responses to developing incidents and are key to reducing crime, injury and security incidents.

III. Documentation and analysis of events and incidents identify common causes, trends, and patterns that predict and prevent crime, injury, and other incidents.

IV. Training of #ORG# security staff and #ORG# staff is critical to their performance. Staff are trained to recognize and to immediately report potential and actual security incidents. Security staff in sensitive areas are trained about protective measures designed for those areas and their responsibilities to assist in protection of patients, staff, visitors, and property.

V. Policies and procedures are in place to guide the Police or Security Department Officers and staff in their roles and responsibilities for responding to a variety of emergencies and security incidents.

VI. There are protocols and education programs that address a safe, secure, and therapeutic environment.

VII. The security program establishes processes to reduce the occurrence, the probability, and the effects of person-to-person violence.

VIII. The goals of the secure environment plan are:
   A. Comply with accepted standards of security.
   B. Provide a safe, secure, and therapeutic environment for patients, staff, and visitors.
   C. Integrate security practices into daily operations.
   D. Identify opportunities to improve performance.

IX. The organization's responsibilities for the secure environment plan are:
   A. The Administrator receives regular reports on activities of the security program from the Safety Officer. The Administrator reviews reports and, as appropriate, communicates security related concerns about identified issues and regulatory compliance.
   B. The Administrator reviews reports and, as necessary, communicates concerns about key issues and regulatory compliance to appropriate departments, services, and staff. The administration collaborates with appropriate departments, services, and staff to establish operating and capital budgets for the security program.
   C. The QSC has responsibility for the identification, collection, and analysis of information regarding security deficiencies, development of plans for improvement, accident and injury prevention and investigation, and emergency response.
   D. Membership on the QSC is by appointment from the Administrator and the BOM and includes representatives from administration, clinical services, and support services. The QSC meets monthly and as necessary to receive reports and to conduct reviews of security issues.
   E. The Administrator authorizes key staff to take immediate and appropriate action in the event of an emergency. An emergency is a situation that poses a threat to life or health, or threatens to damage equipment or buildings.
   F. Department, program and service managers are responsible for orienting new staff members to the department, programs and service-specific security procedures.
   G. Individual staff members are responsible for learning and following job and task specific procedures for security operations. Individual staff members are also responsible for learning and using reporting procedures.

## PROCEDURES:

I. The secure environment management plan identifies how it establishes and maintains security by addressing issues concerning patients (including patient elopement) visitors, personnel, and property.
   A. The Facility's Security Officers are responsible for managing the patrol and response program for security issues and for managing the security for the physical plant and grounds. #ORG# staff is responsible for the day-to-day security of all patients. The Director of Nursing is responsible for establishing policies, procedures, and practices for handling patient elopement. The Facility Engineer is responsible for the maintenance of all physical security devices, such as doors, locks, and lighting with the Safety Officer sharing responsibility for associated activities.
   B. #ORG# addresses security issues, which include access to confidential information, in a manner that allows for full evaluation of issues and the potential impact of changes on the organization. Examples of change might include improvements to the physical plant or modifications to patient care services.

C. The Safety Officer provides the QSC with reports related to security issues.

D. The Information Management Officer is responsible for the security of all electronic data and its confidentiality, access, and integrity.

II. Reporting and investigating all security incidents involving patients, visitors, personnel, or property.

A. The security program uses a variety of reporting methods to document activities. The Safety Officer and Risk Manager share responsibility for managing and investigating incidents. All departments, program and service directors share responsibility for reporting incidents.

B. Reports of staff, patient and visitor incidents are made using the appropriate report forms. The Safety Officer and Risk Manager, and appropriate department, program and service directors review this information. Aggregate information is also reviewed by the QSC.

C. Reports of significant property damage are directed to the Safety Officer and Risk Manager.

D. One of the goals of the reporting process is for the responsible manager to receive #ORG# incident reports as soon as practical after an occurrence. This goal is intended to allow appropriate and timely reporting and follow-up activities as needed.

III. Providing identification, as appropriate, for all patients, visitors, and staff.

A. The Safety Officer is responsible for coordinating the identification program for patients, visitors, and staff. All staff members are required to display identification at all times and to report any persons who do not have proper identification to the Police or Security Department.

B. Leadership, with input from the OSC, develops policies for identification. Policies are reviewed on an as needed basis.

C. Department, program, and service managers are responsible for enforcing identification policies. Violations of identification policies involving patients are reported to Department, program, and service staff.

D. The Finance Officer issues and recovers contractor/vendor identification.

IV. Controlling access to and egress from sensitive areas, as determined by #ORG#.

A. The Safety Officer is responsible for managing the program to identify sensitive areas. The Administrator is responsible for approving those areas that have been designated as sensitive.

B. The following areas have been designated as sensitive areas:
   1. Medical Records Storage and Offices
   2. Pharmacy and Medical Supply Storage
   3. Pediatric Areas
   4. Operating Room and PACU
   5. Public and Employee Parking Areas
   6. Cashier Area
   7. Patient Care Areas
   8. Computer Systems Room
   9. Telephone Systems Room

C. Staff is instructed about all areas of #ORG# that have been designated as security sensitive. Staff assigned to work in sensitive areas receives service-specific education that focuses on special precautions and responses pertaining to their area.

V. Leadership's designation of personnel responsible for developing, implementing, and monitoring the security management plan.

A. The Administrator manages the process for designating the personnel responsible for developing, implementing, and monitoring the security program.

B. By designation, the Safety Officer is responsible for managing the security program.

C. The Safety Officer reviews changes to regulations; assesses needed changes to security equipment; and performs activities essential to implement the security program. The Safety Officer shares responsibility with department, program, and service Directors for security activities associated with risk management and emergency management.

VI. The annual evaluation of the security management plan's objectives, scope, performance, and effectiveness.
   A. The QSC Chair has overall responsibility for coordinating the annual evaluation of each of the seven functions associated with Management of the Environment of Care. The Safety Officer is responsible for completing the annual evaluation of the security program. An evaluation of the program's objectives, scope, performance, effectiveness, and the Secure Environment Plan is included in each annual evaluation.
   B. In the completion of the annual evaluation, the Safety Officer utilizes a variety of source documents such as policy review and evaluation, incident report summaries, risk assessment activities, meeting minutes, and statistical information summaries. In addition, other relevant sources of information are used for the annual evaluation, such as results of monitoring studies, reports from accrediting and certification agencies, and goals and objectives. The annual evaluation of the security program is used to further develop educational programs, policies, and performance monitoring and performance improvement.
   C. The annual evaluation is reviewed and approved by the QSC. The annual evaluation is then presented to the Administrator, and Board of Managers. Minutes or other means of communications from the Administrator, and Board of Managers are received, reviewed, and acted upon by the QSC.

VII. Emergency security procedures.
   A. The Safety Officer has overall responsibility for coordinating activities related to establishing and maintaining emergency procedures for the security program. The Director of each department program and service designated as a security sensitive area is responsible for coordinating the activities with regard to establishing and maintaining emergency procedures. Each Director is responsible for developing emergency security procedures that consider the needs of patients and other departments, programs and services, especially clinical areas.
   B. The Safety Officer and the QSC each have responsibility for reviewing emergency security procedures related to the operations of sensitive areas.
   C. Emergency security procedures contain specific information related to actions to be taken in the event of a security incident; handling of civil disturbances; situations involving VIPs and the media; and provisions for providing additional personnel to control traffic during emergencies.
   D. Each Director is required to structure their department, program, and service security procedures to emphasize that procedures provide personnel with essential information needed during an emergency.
   E. Each Director is responsible for maintaining copies of departmental, program, or service emergency procedures in a location accessible to their staff for reference during an emergency. Directors are responsible for providing their staff with orientation to emergency procedures that relate to their jobs. Additional department level training is provided on an annual basis as part of the continuing education program and when emergency procedures are revised.
   F. Directors are responsible for maintaining the emergency procedures. Each Director is responsible for reviewing emergency security procedures and updating as necessary. The Safety Officer is responsible for coordinating the review program for emergency procedures and for communicating information related to the findings of the review to the QSC.

VIII. Providing vehicular access to urgent care areas.
   A. The Safety Officer is responsible for designating and managing the program that provides vehicular access to urgent, regulated, and emergency equipment areas.
   B. Vehicular access is enforced by the Police Department. The Safety Officer is responsible for reviewing practices to determine if changes are needed.

IX. A security orientation and education program that addresses security issues.
   A. Administration has overall responsibility for organizing the orientation and education program for all environments of care processes. Department, program, and service directors are responsible for assuring the secure environment program orientation and education is implemented.
   B. Administration is responsible for conducting the general orientation program with current information on general safety processes to new staff members as soon as possible but within thirty (30) days of employment. Every new staff member participates in a general orientation program that includes information related to the secure environment. Critical safety information is provided prior to staff being allowed to

work independently. Attendance is recorded for each new staff member who completes the general orientation program and these records are kept in administration.

C.  Each department, program, and service director is responsible for providing their new staff members with secure environment orientation specific to their department, program, and service. The goal of these orientation programs is to provide new staff members with current job specific safety and hazard information.

D.  All staff members of #ORG# participate in mandatory continuing education at least once each year, which includes information specific to the secure environment program. This requirement may be satisfied through completion of a self-learning packet provided by administration.

E.  The Safety Officer reports information on orientation and continuing education data during the reporting period to the QSC.

X.  Ongoing monitoring of performance regarding actual or potential risks.

A.  The Safety Officer through the QSC has overall responsibility for coordinating the ongoing performance monitoring and the performance improvement monitoring for all functions associated with Management of the Environment of Care. The QSC is responsible for all monitoring associated with the security program.

B.  The intent of establishing performance monitoring is to improve the security program through objective measures of demonstrated performance. The results of measurement are reported through appropriate channels including #ORG#'s leadership and when appropriate to relevant components of #ORG# wide patient safety program. Performance improvement is an important aspect of the Secure Environment Plan. Ongoing performance monitoring serves as an indicator of continued effectiveness of the security program and is a mechanism to identify performance improvement opportunities.

**REFERENCES:**

**FORMS:**

**EQUIPMENT:**

**APPROVALS:**

| NAME | TITLE | DATE |
| --- | --- | --- |
| #APPROVER# | #APPRTITLE# | #APPRDATE# |

**Policy and Procedures**

Policy#:  #POLNUM#
Location:  #LOC#
Originating Department:  #ORIGDEPT#
Effective Date:  #EFFDATE#
Expiration Date:  #EXPDATE#

## TITLE: SINGLE AND MULTI-DOSE MEDICATION CONTAINERS

### POLICY STATEMENT:

It is the policy of #ORG# that standard guidelines are maintained for the use of single- and multi-dose containers.

### INTENT AND SCOPE:

This policy affects all Medical Staff and employees (contract and non-contract) of #ORG#.

### DEFINITIONS:

### GENERAL INFORMATION:

I. All parenteral drugs available from the manufacturers have indications on their label as to the intended use, i.e., single use or multiple uses.

II. Drugs labeled for single use are discarded immediately after single use.

III. Unless the manufacturer designates a shorter expiration date, multi-dose containers with antimicrobial preservatives dispensed from pharmacy expire twentyeight (28) days after opening date, provided:
   A.   There is no obvious contamination.
   B.   Aseptic technique has been followed when withdrawing the medication.
   C.   The MDV is stored according to the manufacturer.

IV. MDVs are discarded:
   A.   When empty
   B.   When suspected contamination occurs
   C.   When contamination/particulates are visible
   D.   When the vial has reached twentyeight (28) days from opening or when the manufacturer's established expiration date is reached, as long as the MDV has been stored according to manufacturer's instructions

V. Sterile solutions of normal saline and water with reclosable tops used for irrigation are discarded twentyfour (24) hours after the container is opened.

**Exception:** Refrigerated Double Antibiotic (DAB) solution has a seven (7) day expiration date from the day of opening.

   A.   Urine and blood test products such as Dextrostix, Acetest tablets, and Tes-tape rolls are discarded ninety (90) days after being opened or the manufacturer's expiration date, whichever comes first.
   B.   Expiration Date or Beyond Use Date (BUD) and initials of user must be written on the original container when the multi-dose vial is first opened or entered (i.e., needle-punctured).

VI. Multi-dose parenterals used on individual patients are labeled with the patient's name and the expiration date of the opened product. These are stored in the patient's medication drawer if refrigeration is not required. Multi-dose containers, requiring refrigeration, follow the manufacturer's recommendation for expiration after opening.

## PROCEDURES:

I. The healthcare provider performs hand hygiene before handling the multi-dose vial.

II. A new, sterile needle and syringe are used each time medication is drawn from the vial.

III. The integrity of the stopper is checked and the medication in the vial checked for any particulate matter prior to use. If the integrity of the stopper or sterility of the medication is in question, discard the vial.

IV. The stopper or vial gum is swabbed with 70% alcohol before each puncture.

V. Avoid touching the stopper of the MDV during this process.

VI. Pharmacy professionals verify that MDVs are stored and labeled correctly when inspecting medication storage areas.

## REFERENCES:

## FORMS:

## EQUIPMENT:

## APPROVALS:

| NAME | TITLE | DATE |
|------|-------|------|
| #APPROVER# | #APPRTITLE# | #APPRDATE# |

**Policy and Procedures**

Policy#: #POLNUM#
Location: #LOC#
Originating Department: #ORIGDEPT#
Effective Date: #EFFDATE#
Expiration Date: #EXPDATE#

## TITLE: SURGICAL SITE IDENTIFICATION

**POLICY STATEMENT:**

It is the policy of #ORG# that Attending Physicians are ultimately responsible for mandating that the correct procedure be performed for the patient. All procedures follow the informed consent policy of #ORG#.

**INTENT AND SCOPE:**

This policy is intended to establish and communicate a standard methodology for mandating that surgery be performed on the intended site/side. This policy affects all medical staff.

To mandate that the correct procedure (including site/side) is marked and performed on the correct patient, on all operative and other invasive procedures that expose patients to more than minimal risk.

**DEFINITIONS:**

**GENERAL INFORMATION:**

I. Marking the site/side is required for all procedures except those done through or immediately adjacent to a natural body orifice i.e., Tonsillectomy, Hemorrhoidectomy, procedures on genitalia or other situations in which marking the site/side would be impossible or technically impractical. Marking the site/side is still required for cases involving right/left distinction, multiple structures (such as fingers and toes) or levels (such as spine).

II. #ORG# does not require the site/side to be marked for other types of procedures, including mid-line sternotomies for Open Heart Surgery, Cesarean Sections, Laparotomy, teeth extraction (but indicate tooth name(s) on documentation or mark the operative tooth (teeth) on the dental radiographic or dental diagram), Laparoscopy and interventional procedures for which the site/side of insertion is not pre-determined, such as Cardiac Catherization procedures. In the above scenarios, all other procedures described below must still be performed.

III. Induction of anesthesia is not initiated prior to completion of the Surgical Site Identification or the Procedural Checklist.

IV. The skin is to be marked, even if patient has a dressing, cast, or traction device.

V. This policy is enforced for all patients. No exceptions are tolerated except, in a true loss of life or limb emergency. The attending physician/resident documents this exception in the progress note.

**PROCEDURES:**

I. The patient is identified by asking the patient to confirm full name, date-of-birth, medical record number, or social security number, and procedure. The Registered Nurse during the pre-operative assessment asks the patient to point to the correct body site/side for the procedure except in the posterior spinal region, where it is physically impossible for the patient to reach. The word "yes" is placed by the nurse on the appropriate site/side and placed on marking form. If bilateral sites/sides all is marked with word "yes" and all placed on marking site form.

II. In addition to pre-operative skin marking of the general spinal region, special intraoperative radiographic techniques are used for marking the exact vertebral level.

III. The mark is positioned to be visible after the patient is prepped and draped. The mark is on or near the incision site; the mark is visible and sufficiently permanent to remain visible following skin preparation and draping.

IV. The letter "X" is never used to mark a site/side.

V. The Registered Nurse verifies the site/side that the patient identified, and it must agree with the scheduled procedure, the informed Consent, the H&P and the Radiology report (if applicable), or other test results (if applicable).
   A. If there is a discrepancy between both the consent and the patient verification, the surgeon/physician is called and the procedure is delayed until the surgeon/physician arrives and verifies the correct site/side.
   B. If the patient is unable to communicate or is unreliable or unresponsive the health care members attempt to complete site/side verification with a family member or significant other. If unavailable, they rely on physical assessment and clinical documentation. The Registered Nurse calls the surgeon/physician and the surgeon/physician arrives to verify the correct site/side prior to pre-medication and mark the site/side.
   C. No patient is taken to the surgical suite or procedural area until all verifications match.

VI. The Attending Physician who is performing the procedure verifies the site/side with the patient against the consent before sedative drugs are administered. For the incompetent and/or incapacitated patient, a family member, friend or a responsible adult who is not a member of the operative team represents the patient in verifying the operative site/side against the consent.

VII. The Attending Physician marks the operative/procedure site/side with his/her initials in a way that cannot be overlooked and in a manner that is clearly incorrect if transferred to another site/side prior to surgery. This is done pre-operative when the physician evaluates and speaks to the patient.

VIII. No sedation is administered until site/side verification process has been completed. The preparation of the competent patient for procedure is complete prior to receiving medication that could impair cognitive function. This does not eliminate the administration of analgesics if the patient's clinical condition warrants.

IX. Upon interview by the operating room nurse, the patient will again point to the correct site/side and this site/side is placed on the board in the operating room.

X. The Surgical Team prior to anesthetizing the patient and utilizing the holding room, the first verification check that is actively communicated; that the correct patient, the correct procedure including site/side is marked and match, the consent, H&P, and radiological films (if applicable) upon presenting to OR suite. The team verifies that any implants, equipment or any other special requirements are present. The surgeon confirms all bilateral site/sides are marked accurately. All site/sides are written on the dry erase boards in the operating room by the OR nurse.

XI. Immediately prior to initiating the incision, the practitioner and members of the team conducts a "time out" to review the following with the physician:
   A. Correct implants; type and size
   B. Specialized equipment or other specialized requirement needs
   C. Correct site/side
   D. Correct procedure
   E. Correct position
   F. Correct medications or other biologicals

This "time out" is documented on the Operative/Procedural record and Surgical Identification Checklist by the Registered Nurse in surgery/procedural area.

XII. If there are any disagreements about the above criteria, (including instances where patients refuse site/side marking) the procedure does not proceed until the disagreements are reconciled and documented on the Surgical Site Identification Checklist by the Registered Nurse.

XIII. Once the patient is positioned for the surgery/procedure, the Physician makes the incision through or adjacent to his/her initials.

XIV. Site/side verification is documented on the Surgical Site Identification Checklist.

This policy is enforced for all patients. No exceptions are tolerated, except in a true Life and Death emergency. The Attending Physician documents this exception in the progress note.

**REFERENCES:**

**FORMS:**

**EQUIPMENT:**

**APPROVALS:**

| NAME | TITLE | DATE |
|------|-------|------|
| #APPROVER# | #APPRTITLE# | #APPRDATE# |

**Policy and Procedures**

Policy#:  #POLNUM#
Location:  #LOC#
Originating Department:  #ORIGDEPT#
Effective Date:  #EFFDATE#
Expiration Date:  #EXPDATE#

## TITLE: UTILITIES MANAGEMENT PLAN

### POLICY STATEMENT:

It is the policy of #ORG# that #ORG# Utilities Management Program's scope: Maintains a utility systems management program which promotes a safe, controlled and comfortable environment of care; minimizes the risks of utility failures; and ensures operational reliability of utility systems.

### INTENT AND SCOPE:

This policy is intended to establish and communicate a standard approach to maintain a utility system management program. This policy applies to all Medical Staff and employees (contract and non-contract) of #ORG#. The scope includes electrical, medical air, medical vacuum, medical gas, water, emergency generator, boiler and steam, sewer, natural gas, heating, ventilation and air conditioning, vertical transport, communication, and Life Safety Systems.

### OBJECTIVES:

1. Ensure testing, inspection and maintenance of critical components of utilities.
2. Provides utility system operational plans, maps, and labeling controls.
3. Investigation of utility failures with creation and implementation of corrective action plans.
4. Prevention and reduction of utility incidents which result in unplanned failures and/or interruption of services.
5. Minimize risks through appropriate assessments.
6. Labeling of controls so that complete or partial emergency shutdown can occur.
7. Ensure that the utility systems are operational, reliable and functional.
8. Provide an inventory of all equipment and utilities.
9. Educate staff, physicians, practitioners and contract workers.
10. Develop and implement emergency procedures.

### DEFINITIONS:

### GENERAL INFORMATION:

I. Utility systems management is a function and set of activities focused on meeting #ORG#'s utility systems needs.

II. Assess and minimize risks of utility failures and ensure operational reliability of systems is included in #ORG#'s Utilities Management Plan.

III. The Utilities Management Program is designed to be aware and reduce organizational-acquired illness, by monitoring indoor air quality, preventative maintenance of Heating, Ventilation, & Air Conditioning (HVAC) systems, infection control procedures, water treatment to minimize legionella or other anaerobic growth in the air systems and following guidelines for ventilation requirement for the facility.

IV. The Utilities Management Program is designed to confirm operational reliability, assess risks, respond to failures, and to train users and operators of the utility system components, thus promoting a safe, controlled and comfortable environment.

V. There is a comprehensive preventative maintenance program which includes a written testing and maintenance program for all utility components included in the program at established intervals of annually or more if needed. It is the responsibility of Facility Operations to keep the preventative maintenance program accurate and ongoing.

VI. Criteria are established for identifying, evaluating and taking inventory of critical operating components of systems to be included in the utility management program which include equipment that meet the following criteria:
   A. Equipment maintains the climatic environment in patient care areas.
   B. Equipment that constitutes a risk to patient life support upon failure.
   C. Equipment is a part of a building system which is used for infection control.
   D. Equipment that is part of the communication system which may affect the patient or the patient care environment.
   E. Equipment is an auxiliary or ancillary part of a system control or interface to patient care environment, life support or infection control.

VII. The following systems are included in the Utilities Management Program:
   A. Electrical distribution system,
   B. Emergency power system,
   C. Vertical transport (elevators),
   D. Heating, ventilation and air conditioning systems,
   E. Chiller systems,
   F. Cooling tower operations and treatment,
   G. Pneumatic tube system,
   H. Fuel distribution system,
   I. Natural gas system,
   J. Plumbing and water delivery systems,
   K. Boilers, condensate and steam delivery systems,
   L. Medical gas distribution,
   M. Medical and surgical vacuum and air delivery systems,
   N. Sewage Removal systems,
   O. Communications systems,
   P. Infant Abduction Prevention System,
   Q. Alarm System.

VIII. A scheduled maintenance system exists which is used to schedule, monitor and document the testing and maintenance of each utility system at predetermined levels.

IX. A comprehensive preventative maintenance program includes written testing and maintenance programs for all utility components help to ensure reliability to minimize risks and reduce failures of utilities systems. It is the responsibility of Facility Operations to keep the preventative maintenance program accurate and ongoing at the established intervals. Preventative maintenance will be performed annually or more if needed.

X. There are drawings mapping the distribution of utility systems which indicate the controls for partial or complete shutdown of each utility system. All emergency shutoff controls of for the utility systems components shall be labeled clearly, visibly and permanently throughout the facility. See Utility Systems Shutdown, Utility Systems Record Drawings, Utilities Management Emergency Shutoff Labels Policy and Sample Labels available in Facility Operations/Bio Med and administration.

XI. Investigation and reporting incidents and corrective actions of utility systems management problems, failures and user errors:
   A. A Utility System Failure Report is completed for any problem, failure or user error of a vital or essential system.
   B. Facility Operations responds and correct all identified problems within the scope of their operations in a timely manner. Evidence of the actions taken to resolve identified problems can be located in the Facility Operations Daily Log, the completed work orders file, the utilities management failure, user error log and additionally the problem resolution log.

XII. #ORG#'s Facility Operations personnel are required to attend an orientation upon hire and regularly scheduled in-services that specifically address utility systems capabilities, limitations, special applications, emergency procedure if failure occurs, maintenance responsibilities, the location and instruction on use of emergency shut off controls and reporting procedures for utility systems problems, failures and user errors. All user/maintainers of equipment are tested for competency according to the components of their job specifications.

## PROCEDURES:

## PERFORMANCE STANDARDS

I. There is a systematic, interdisciplinary and interdepartmental process to monitor, assess, evaluate and improve the quality of services provided by the Facility Operations department.
  A. Percent of staff able to demonstrate their knowledge and skill of their role and expected participation in the Utilities Management Program.
  B. Number of utility incidents report.
  C. Number of user errors reported.
  D. Number of utility failures or interrupts.

II. Written procedures are developed that specify the action to be taken during the failure of major utility services. Emergency procedures include:
  A. Procedures to follow when a utility system malfunctions;
  B. Alternate sources of essential utilities;
  C. Shut off procedures and controls of malfunctioning systems;
  D. Procedures for notifying personnel in the affected areas;
  E. How to obtain repair services; and
  F. Procedures to perform emergency clinical interventions.

III. The written procedures include a call system for summoning essential personnel and outside assistance when required.

The annual evaluation of the Utilities Management Program includes a review of the scope and objectives by Facility Operations. The performance and effectiveness of the Utilities Management Program is reviewed by the Quality and Safety Committee, and Administration and is approved annually by the Board of Managers.

## REFERENCES:

## FORMS:

## EQUIPMENT:

## APPROVALS:

| NAME | TITLE | DATE |
|------|-------|------|
| #APPROVER# | #APPRTITLE# | #APPRDATE# |

**Policy and Procedures**

Policy#: #POLNUM#
Location: #LOC#
Originating Department: #ORIGDEPT#
Effective Date: #EFFDATE#
Expiration Date: #EXPDATE#

## TITLE: WASTE DISPOSAL

### POLICY STATEMENT:

It is the policy of #ORG# that all special waste is disposed of in a manner that meets or exceeds the standards of local, state and federal licensing, regulatory and accrediting agencies.

### INTENT AND SCOPE:

This policy is intended to provide compliance with all guidelines.

This policy applies to all Health Care Workers (HCW) of #ORG# (contract and non-contract), Medical staff and volunteers as defined by the Center for Disease Control and Prevention.

### DEFINITIONS:

I. HCW: "HCWs" refers to all the paid and unpaid persons working in health care settings. This may include, but is not limited to, physicians, nurses, aides, dental workers, technicians, workers in laboratories and morgues, emergency medical service (EMS) personnel, students, part-time personnel, temporary staff not employed by the health-care facility, and persons not involved directly in patient care but who are potentially at risk for occupational exposure (e.g., volunteer workers, nutrition, housekeeping, maintenance, clerical, and janitorial staff).

II. Personal Protective Equipment (PPE) is equipment and supplies designed to protect the HCW from exposure to blood and body fluids including, but not limited to, protective eyewear, gowns, aprons, masks, and particulate respirators.

III. "Special waste" refers to infectious waste from health care related facilities; solid waste which, if improperly treated or handled, may transmit an infectious disease(s) and which is comprised of the following:
   A. Blood and blood products—All waste bulk human blood, serum, plasma and other blood components. Bulk blood or body fluids means a volume of 100 ml or more and/or items that are saturated **and** drip without compression.
   B. Microbiological Waste—Microbiological waste includes:
      1. Cultures and stocks of infectious agents and associated biological,
      2. Cultures of specimens from medical and pathological laboratories,
      3. Discarded live and attenuated vaccines,
      4. Disposable culture dishes, and
      5. Disposable devices used to transfer, inoculate and mix cultures.
   C. Pathological waste—Pathological waste includes but is not limited to:
      1. Human materials removed during surgery, labor and delivery, autopsy, or biopsy, including; body parts, tissues, fetuses, organs, and bulk blood and body fluids.
      2. Products of spontaneous or induced human abortions regardless of the period of gestation.
      3. Laboratory specimens of blood and tissue after completion of laboratory examination.
      4. Anatomical remains.

    D.   Sharps—Sharps includes the following materials when contaminated:
1. Hypodermic needles;
2. Hypodermic syringes with attached needles;
3. Scalpel blades;
4. Razor blades and disposal razors used in surgery, labor and delivery, or other medical procedures;
5. Pasteur pipettes; and
6. Broken glass from laboratories.

IV. Regulated Waste—"Means liquid or semi-liquid blood or other potentially infectious materials: contaminated items that would release blood or other potentially infectious materials in a liquid or semi-liquid state if compressed; items that are caked with dried blood or other potentially infectious materials and are capable of releasing these materials during handling; contaminated sharps; and pathological and microbiological wastes containing blood or other potentially infectious materials."

V. Ordinary waste—Waste such as paper that does not come in contact with blood and/or body fluids.

VI. Hazardous waste—A waste is hazardous if it exhibits one of the following characteristics: ignitability, corrosiveness, reactivity or toxicity.

## General Information:

I. Ordinary waste, hazardous waste, infectious waste, sharp instruments, sharp disposable items, and sharps containers must be separated at their source and transported to the appropriate processing area.

II. This policy does not address the handling or disposal of microbiological, pathological or hazardous waste. Reference the following policies and/or departments regarding the handling and disposal of hazardous waste:
A. Hazardous Materials Management
B. Total Waste Management
C. Antineoplastic Agent Administration
D. Handling of Cytotoxic Agents
E. Management of Radioactive Materials

III. Container requirements for the disposal of contaminated sharp instruments and contaminated disposable sharp items must be:
A. Closable, puncture resistant, leak-proof on all sides and bottom.
B. Red in color and/or labeled with the biohazard symbol.

IV. Sharps containers are handled in the following manner:
A. Maintained in an upright position during transport.
B. Transported in a manner to have minimal contact with personnel.

V. Container requirements for infectious/regulated waste (red plastic bags meet these requirements):
A. Closable.
B. Constructed to contain all contents and prevent leakage of fluids during handling, storage, transportation or shipping.
C. Labeled with the biohazard symbol and/or red in color.

VI. Infectious/regulated waste (non-sharp) are placed in red bags and disposed of in the following manner:
A. Are tied off to securely close the bag prior to disposal.
B. Are disposed of promptly.
C. If the integrity of the bag is breached or the outside of the bag is contaminated, then the bag is placed in a secondary container or bag, which meets the container requirements.

VII. Ordinary waste is discarded in regular plastic bags.

**PROCEDURES:**

I. Disposal of Waste

   A.  Disposal Containers

      1.  Ordinary waste is discarded into a waste container lined with a regular plastic bag.

      2.  Infectious/regulated waste is discarded in compliance with Isolation Precautions #ORG# Policy

   B.  Environmental Services:

      1.  Collects ordinary waste from each area on a regularly scheduled basis. Takes to a compactor where it is held for pick up. No further processing is necessary before collection by a contracted disposal service.

      2.  Collects red bag waste separately from ordinary waste for transport to the processing area.

      3.  Checks waste containers and discards waste appropriately as needed.

      4.  Ties the top of the bag or otherwise securely closes the bag prior to disposal, to prevent spillage during handling, transporting and/or shipping.

      5.  Transports waste for disposal from the point of collection by the most direct route for disposal.

      6.  Maintains a log of records for special waste processed on site. Records must be maintained for a period of three years.

      7.  Routinely checks sharps containers and replaces containers that are full.

      8.  Transports the filled, closed sharps containers in an upright position to the processing area.

      9.  Operators are provided appropriate PPE to prevent skin and clothing contamination.

    10.  Only trained Environmental Services staff load, clean and dispose of special waste.

          Handling or Treatment of Special Waste on Site

          Develop a written procedure for the operation and testing of equipment used.

          Operator maintains a log of written records of special waste treated on site, recording the following:

            –  The date of treatment;

            –  The amount of waste treated;

            –  The method/conditions of treatment;

            –  The name (printed) and initials of the person(s) performing treatment.

          Handling or Treatment of Special Waste off Site

          Special Waste is not allowed to accumulate more than 60 cubic yards prior to the decision being made for shipping off site.

   C.  Requirements for shipment of untreated special waste off site:

      1.  Environmental Services supervisor is responsible for contacting the company with whom #ORG# has a back-up contract to transport special waste.

      2.  Environmental Services supervisor actively supervises persons (Environmental Services staff and waste disposal contractor) responsible for the packaging of special waste to be shipped off site.

      3.  Untreated special waste which is shipped off site for treatment or disposal must be identified and packaged per the following requirements, with supplies provided by the waste disposal contractor:

         a.  Special waste, other than sharps, are placed in a plastic bag, which meets the requirements of the American Society of Testing and Measurement, Standard No. D 1709-85 using a 165-gram dart. If empty containers that hold free liquids are placed into the bag, 1 cup of absorbent material for each six cubic feet, or fraction thereof, of bag volume is placed in the bottom of the bag.

         b.  The bag containing special wastes is placed in a rigid container, which is constructed of a material that meets or exceeds the strength of 200 pound, C-Flute board.

         c.  If the waste contains free liquids in containers, the plastic bag and/or the rigid container must contain absorbent material sufficient to absorb 15% of the volume of free liquids placed in the bag.

         d.  The outer container is conspicuously marked with a warning legend that appears in English and Spanish, along with the international symbol for biohazard material. The warning appears on the sides of the container, twice in English and Spanish. The wording of the warning reads as follows: CAUTION, contains medical waste which may be biohazardous and PRECAUCION contiene desperdicios medicos que pueden ser peligroso para el cuerpo humano.

         e.  Each outer container is labeled with #ORG#'s name, address, and either the date of shipment or an identification number for the shipment.

         f.  The transporter affixes to each container a label that provides the name, address telephone number, and state registration number of the transporter. This information may be printed on the container.

g. The printing on labels required in d, e, and f above is done in indelible ink with letters at least 0.5 inch in height. A single label may be used, if a single label is used the transporter insures the label is affixed to or printed on the container.

h. Sharps are placed in a marked, puncture-resistant rigid container designed for sharps. The sharps container may be placed in a plastic bag (which meets the requirements described in #1 of this section). The bag is placed in a rigid container (which meets the requirements described in #2 of this section).

4. Shipments of untreated special wastes are released to the #STATE# Department of Health to transport special wastes from health care facilities. Release only to a transporter who is registered for the untreated waste to unregistered transporters is a violation of #STATE# Water Commissions' Solid Waste Management Rules.

5. Environmental Services staff:

a. Obtain a signed receipt for each shipment of regulated medical.

b. Maintain a file of receipts/manifests for shipments of special waste for a period of three years following the date of shipment. The file contains the receipt signed at the time of shipping and the receipt/manifest returned by the transporter with the date and time of incineration and the signature of the person responsible for the incineration. The three-year period for record maintenance may be extended for investigative purposes or in case of enforcement action. Failure to maintain the file of receipts in an orderly fashion, destruction of receipts prior to the end of the specified time or destruction of receipts prior to the expiration of an extended retention time is a violation of the standards.

c. Make the file of receipts for shipments of special wastes available for inspection during normal business hours without notice. Refusal is a violation of the standards.

**REFERENCES:**

**FORMS:**

**EQUIPMENT:**

**APPROVALS:**

| NAME | TITLE | DATE |
| --- | --- | --- |
| #APPROVER# | #APPRTITLE# | #APPRDATE# |

# References

1. Accreditation Association for Ambulatory Health Care (AAAHC)
   http://www.aaahc.org/eweb/StartPage.aspx

2. American Academy of Sleep Medicine
   http://www.aasmnet.org/

3. American Association for Accreditation of Ambulatory Surgery Facilities, Inc.
   http://www.aaaasf.org/consumers.php

4. American College of Radiology
   http://www.acr.org/

5. American College of Surgeons- Bariatric Surgery Center Network
   http://www.acsbscn.org/Public/AboutBSCN.aspx

6. American Commission for Health Care
   http://www.achc.org/

7. American Health Information Management Association (AHIMA).
   www.ahima.org
   Joint Commission Resources. *A Patient Safety Handbook for Ambulatory Care Providers.* Oakbrook Terrace, IL: Joint Commission Resources, 2009.

8. Ambulatory Surgery Center Association
   http://www.ascassociation.org/businessdirectory/accreditationorganizations

9. APIC. "Ambulatory Care." Chap. 49 in *APIC Text of Infection and Control and Epidemiology,* 3rd ed. Washington D.C.: APIC, 2009.

10. Bennet, Gail. *Infection Prevention Manual for Ambulatory Care.* Rome, GA: ICP Associates Inc., 2009.

11. CDC, APIC, SHEA. *How-to Guide: Improving Hand Hygiene: A Guide for Improving Practices among Health Care Workers.*
    http://www.shea-online.org/Assets/files/IHI_Hand_Hygiene.pdf

12. Center for Medicare and Medicaid Services (CMS)
    http://www.cms.gov/CFCsAndCoPs/

13. Community Health Accreditation Program
    http://www.chapinc.org/

14. Conditions for Coverage for ASC's (CMS)
    http://www.cms.gov/CFCsAndCoPs/16_ASC.asp#TopOfPage

15. Guinane, Carole and Noreen Davis. *Improving Quality in Outpatient Services.* Boca Raton, FL: Productivity Press, 2011.

16. Hand Hygiene in Healthcare Settings. Center for Disease Control.
    http://www.cdc.gov/handhygiene/

17. Healthcare Facilities Accreditation Program
    http://www.hfap.org/AccreditationPrograms/amb_care.aspx

18. Infection Prevention and Control in Healthcare. World Health Organization.
    http://www.who.int/csr/bioriskreduction/infection_control/en/

19. Institute for Medical Quality
    http://www.imq.org/programs/ambulatory-care/accreditation/

20. National Accreditation Program for Breast Centers–administered by the American College of Surgeons
    http://accreditedbreastcenters.org/standards/standards.html

21. National Committee for Quality Assurance (NCQA)
    http://www.ncqa.org/

22. Physician Office Laboratory Accreditation (COLA)
    http://www.cola.org/pol.html

23. Surgical Review Corporation COE for freestanding outpatient bariatric programs
    http://www.surgicalreview.org/pcoe/free_standing/freestanding.aspx

24. The Joint Commission (TJC). 2011. Comprehensive Accreditation Manual. CAMAC for Ambulatory Care.

25. The Joint Commission Ambulatory Care Accreditation (TJC)
    http://www.jointcommission.org/AccreditationPrograms/AmbulatoryCare

26. The Joint Commission Disease Certification Programs
    http://www.jointcommission.org/CertificationPrograms/Disease-SpecificCare/HTBC/dsc_eligibility.htm

27. Urgent Care Association of America
    http://www.ucaoa.org/

28. Urgent Care Center Accreditation (UCCA) through the American Academy of Urgent Care Medicine
    http://www.aaucm.org

29. World Health Organization. 2005. WHO Guidelines on Hand Hygiene in Health Care: A Summary. Clean Hands are Safer Hands.
    http://www.who.int/gpsc/tools/en/

# Index

For Product Safety Concerns and Information please contact our EU
representative GPSR@taylorandfrancis.com Taylor & Francis Verlag GmbH,
Kaufingerstraße 24, 80331 München, Germany

Printed and bound by CPI Group (UK) Ltd, Croydon, CR0 4YY
01/05/2025
01858599-0001